PRECISION MANUFACTURING COSTING

E. RALPH SIMS, JR., P.E.
*Associate Professor of Industrial
 and Systems Engineering*
Ohio University
Athens, Ohio

Founder
The Sims Consulting Group, Inc.
Lancaster, Ohio

Marcel Dekker, Inc.   New York • Basel • Hong Kong

Library of Congress Cataloging-in-Publication Data

Sims, E. Ralph, Jr.
  Precision manufacturing costing / E. Ralph Sims, Jr.
     p. cm. -- (Cost engineering ; 23)
  Includes bibliographical references and index.
  ISBN 0-8247-9083-9 (alk. paper)
  1. Costs, Industrial. 2. Cost control. I. Title. II. Series:
Cost engineering (Marcel Dekker, Inc.) ; 23.
TS167.S56   1995
658.15'5--dc20   95-3458
                                                            CIP

The publisher offers discounts on this book when ordered in bulk quantities. For more information, write to Special Sales/Professional Marketing at the address below.

This book is printed on acid-free paper.

Copyright © 1995 by E. Ralph Sims, Jr., P.E. All Rights Reserved.

Neither this book nor any part may be reproduced or transmitted in any form or by any means, electronic or mechanical, including photocopying, microfilming, and recording, or by any information storage and retrieval system, without permission in writing from the publisher and the author.

Marcel Dekker, Inc.
270 Madison Avenue, New York, New York 10016

Current printing (last digit):
10 9 8 7 6 5 4 3 2 1

PRINTED IN THE UNITED STATES OF AMERICA

To my wife, Jane, who has supported and shared my professional efforts since 1945. She continues to accept working weekends and late nights as an integral part of our wonderful and enduring marriage. Without her support and understanding, this book could never have been written.

PRECISION MANUFACTURING COSTING

WITHDRAWN

COST ENGINEERING

A Series of Reference Books and Textbooks

Editor

KENNETH K. HUMPHREYS, Ph.D.

Consulting Engineer
Granite Falls, North Carolina

1. Applied Cost Engineering, *Forrest D. Clark and A. B. Lorenzoni*
2. Basic Cost Engineering, *Kenneth K. Humphreys and Sidney Katell*
3. Applied Cost and Schedule Control, *James A. Bent*
4. Cost Engineering Management Techniques, *James H. Black*
5. Manufacturing Cost Engineering Handbook, *edited by Eric M. Malstrom*
6. Project and Cost Engineers' Handbook: Second Edition, Revised and Expanded, *edited by Kenneth K. Humphreys*
7. How to Keep Product Costs in Line, *Nathan Gutman*
8. Applied Cost Engineering: Second Edition, Revised and Expanded, *Forrest D. Clark and A. B. Lorenzoni*
9. Managing the Engineering and Construction of Small Projects: Practical Techniques for Planning, Estimating, Project Control, and Computer Applications, *Richard E. Westney*
10. Basic Cost Engineering: Second Edition, Revised and Expanded, *Kenneth K. Humphreys and Paul Wellman*
11. Cost Engineering in Printed Circuit Board Manufacturing, *Robert P. Hedden*
12. Construction Cost Engineering Handbook, *Anghel Patrascu*
13. Computerized Project Control, *Fulvio Drigani*
14. Cost Analysis for Capital Investment Decisions, *Hans J. Lang*
15. Computer-Organized Cost Engineering, *Gideon Samid*
16. Engineering Project Management, *Frederick L. Blanchard*
17. Computerized Management of Multiple Small Projects: Planning, Task and Resource Scheduling, Estimating, Design Optimization, and Project Control, *Richard E. Westney*
18. Estimating and Costing for the Metal Manufacturing Industries, *Robert C. Creese, M. Adithan, and B. S. Pabla*
19. Project and Cost Engineers' Handbook: Third Edition, Revised and Expanded, *edited by Kenneth K. Humphreys and Lloyd M. English*
20. Hazardous Waste Cost Control, *edited by Richard A. Selg*
21. Construction Materials Management, *George Stukhart*
22. Planning, Estimating, and Control of Chemical Estimation Projects, *Pablo F. Navarrete*
23. Precision Manufacturing Costing, *E. Ralph Sims, Jr.*

Additional Volumes in Preparation

Techniques for Capital Expenditure Analysis, *Henry C. Thorne and Julian A. Piekarski*

Preface

In the modern multinational or global economy, three basic truths have become self-evident. They are:
- Profitable performance requires flexible designs
- Quality and service sell products worldwide
- Competitive pricing requires precision costing

It follows that profitable and competitive performance requires management to design products to meet customer expectations in terms of function, quality, service, and price. Although corporate goals require profitable performance, the public also demands good corporate citizenship and quality products. This means that products must be designed for both performance and disposal while remaining marketable, competitive, and profitable to produce.

Cost Engineering techniques provide some of the tools for achieving these management goals. It is time to rethink our cost measurement and analysis practices and to develop more responsive and precise techniques for determining the cost of manufacturing and the costing of products.

Pricing will always be market dependent and only indirectly related to cost. But the use of *precision costing* will help to define acceptable margins and provide a basis for profitable and competitive marketing decisions and pricing. The objective of this book is to provide a basis for precise cost management in the capital intensive manufacturing of piece part products and in the process industries, where labor and material are often relatively small elements of the cost of the product while burden and overhead are big factors.

The *Precision Engineered Costing System* is a method of costing both piece part (metal, wood, cloth, plastics, ceramics, etc.) and continuous process (chemicals, petroleum, food, cement, paper, etc.) manufacturing operations with particular focus on the cost-finding impact of capital intensive, and high tech operations. These costing concepts are applicable to both serialized high volume production and small lot or job shop operations. The system was conceived by me in 1967 on the basis of practical experience in industry and has since been refined to deal with state-of-art manufacturing practices. It is a composite of generally accepted accounting principles and practices which are reoriented to achieve a total absorption, time use of facility based, true level-by-level cost allocation to work-in-process and finished goods inventory. The time use of facility approach is implemented through the application of a "use rate" or rental approach to the complete absorption of all period and indirect expenses into the product on a work station and equipment operation or use time basis. The system divorces production volume based cost variations and tax based accounting practices from the process to develop a true and more complete measure of manufacturing costs. Each operation assumes its full share of all labor, material, burden, overhead, and general and administrative expense. This results in a level-by-level cost structure which is built in a bottom-up manner based on time use of the manufacturing facilities. The system provides for a true activity based cost system and realistic responsibility accounting. Idle capacity costs and overabsorbed burden and overhead are allocated to the profit and loss accounts, and not to the product. Unused capacity or above plan production can be properly charged, or credited, to the success or failure of marketing or product development, and not to the manufacturing operation.

The presently used "generally accepted accounting principles" were developed over the past two centuries of the Industrial Revolution. They are generally based on labor and materials cost plus a percentage of indirect expense. These indirect costs are applied to products on the basis of production volume. Three endemic problems that exist with this present system have been widely recognized.

1. Generally accepted accounting principles do not provide manufacturing management, design engineering, and product cost estimating with the kind of cost detail and information focus that is required in today's capital intensive, often automated, and complex product manufacturing and warehousing operations.
2. Generally accepted accounting practice generates an erroneous and inverse relationship between computed product cost and current production volume.
3. The typical distribution and accrual type accounting systems often distort operating cost information and performance criteria to accommodate financial policy, management practices, and current tax law requirements.

Preface

Conventional cost accounting techniques often reach erroneous conclusions because they base their analyses on such variable factors as production volume and labor utilization. Many published articles and respected journal papers and many cost accounting professionals have been critical of these traditional practices. (Sutton, 1991; Ashtan and Holmlund, 1991; Seed, 1990, Collins and Werner, 1990; Emig and Mazeffa, 1990; Roth and Borthick, 1989, Kaplan, 1988). Before computers, accountants needed to use percentages, ratios, and averages to generate management reports, summaries, standards, and decision criteria in a timely manner with the available clerical staff. The introduction of computers, data base information management systems, automatic identification, and electronic data reporting has permitted rapid and accurate collection, manipulation, and analysis of information. Data base software programs also permit multiple application of detailed operating data for a variety of management uses and generate a wide variety of reporting formats. It is now possible to have management and tax accounting separated from operational accounting and to use focused analyses of a common data base to produce specialized reports for different audiences.

In today's competitive world, inaccurate and judgmental costing is neither necessary nor acceptable. A new costing approach must be developed to fit the realities of modern industry. Manufacturing and warehousing operations managers and product design and development engineers need precise, volume independent cost data to make sound decisions. However, managers and accountants have not yet developed a precise manufacturing and operating cost philosophy which is useful for both management control and engineering estimating, and is broadly applicable, widely accepted, and logically defensible. In most of the proposed techniques, errors caused by variations in production volume and by distribution allocation of indirect costs continue to distort the results. Existing systems also tend to erroneously blame product cost aberrations on the manufacturing system, even though the variations in production volume which cause them are a result of the success or failure of marketing operations.

In modern high tech industry, most of the expense is attributable to the time use of the facilities and equipment used in the process, and the materials and supplies used in production. In capital intensive operations, direct labor is seldom linearly related to production volume or product cost and often has relatively little impact on total manufacturing cost.

Many businesses still continue to apply the old percentage based, labor oriented, and volume sensitive costing techniques which were developed before the computer made precision costing practical. It is time for a change!

One philosophical change that will be required is the conversion of the costing base from the variables of materials, labor, and volume to the constants of time and the time use of capital facilities in each operation. By allocating all indirect expenses to time use of facilities, indirect and general and administrative expenses (G & A) can be fully absorbed and the correct share of these costs can be precisely assigned to each product, part, and operation, independently from production volume. This technique also permits one to separately account for the over and under use of facilities as a variance on profit or margin, rather than as a false variance on product cost. It is no longer acceptable to achieve "precision by division."

Another philosophical change involves the high return on investment (ROI) hurdle generally used in American industry. As industry becomes more capital intensive, the validity of a three year or shorter ROI becomes increasingly suspect. The most successful modern manufacturing economies in the global market take a much longer-term view of capital recovery. They treat automation as a competitive marketing tool. They deal with the whole enterprise and don't insist on each element of the system being justifiable in isolation. Precision costing of parts, products, and operations helps make high technology decisions more realistic and their results more predictable.

The purpose of this book is to present the philosophy of engineered precision costing based on the time use of facility, total absorption standard costing of operations, parts, and products, and the concept of fully absorbed precision costing and estimating on a level-by-level basis. This philosophy will permit the costing of products and work-in-process prior to and during design, during manufacturing operations, and in the planning and analysis of the manufacturing system's performance. This approach is based on the concepts that:

- Price is independent of cost!
- Cost is based on the time use of facilities and labor, and material.
- Cost is set by product design and manufacturing performance.
- Price is set by the market place and the competition.
- The flow of materials and product through an enterprise is the physical manifestation of the flow of funds through the business.
- Level-by-level control of materials flow and step-by-step cost accumulation is management control of the enterprise.

In order to offer a means for precision costing of manufacturing operations, this book presents some basic, and perhaps to some, heretical management information philosophies and the tools required to apply them. Together, these provide the foundation for a precise manufacturing cost system. They lead to a system of total absorption standard costing which is supported by a matrix part and operation number coding system (language), a level-by-level numerical bill of materials system, and a three-axis matrix information retrieval system. The process collects a proper share of all costs (materials, labor, overhead, G & A, etc.) into the inventory valuation of each part and product at each step in the process, and applies indirect expenses to the product through the medium of a "use rate" based on the time use of facilities and labor for each operation.

Preface

The resulting part, product, or operation cost is independent of the volume of production and can be more precisely applied as a tool in cost engineering, cost estimating, and manufacturing performance evaluation and control.

I hope that the ideas and methods presented in this book will permit management to make more realistic product and marketing decisions in the global economy of the future. I also hope that these techniques will lead to a more realistic and equitable approach to responsibility accounting and activity based costing in both manufacturing and marketing operations. This text is for all of industry and all students of industry. For the practitioner, this text should be treated as a handbook to improve the precision of manufacturing cost management, to measure warehousing operations costs, and to aid the design engineer, manufacturing engineer, and cost estimator in correctly predicting manufacturing performance. At the executive level, this book should help management to understand the errors in the present accounting systems and to improve the competitiveness of their product decisions. At the university level, this book should be the text for the engineering student in cost engineering, engineered costing, production management, and cost estimating. It should be one of the texts used in the business student's study of cost accounting methods and management control techniques.

Many say we need a new approach to cost analysis and accounting in our modern high technology global market. This book offers such a new approach.

E. Ralph Sims, Jr., P.E.

Acknowledgments

This book presents the results of many years of concept development and consulting applications by the author. The coding concepts were first published in 1967 in the author's book *Euphonious Coding* (Management Center of Cambridge). The precision costing philosophy was first publicly presented by the author in his Paper No. 680575 on "Precision Costing of Manufacturing Operations" at the Society of Automative Engineers' (SAE) National Combined Farm, Construction and Industrial Machinery, Powerplant, and Transportation Meetings in Milwaukee, Wisconsin, September 9–12, 1968.

Many of the book's illustrations are included courtesy of Edward J. Phillips, P.E., C.M.C., President of The Sims Consulting Group, Inc., of Lancaster, Ohio. For their use I say thanks!

I would also like to commend and thank three men who tested specific techniques upon which this book is based. They are:

> Anas Alsawaf, one of my graduate students at Ohio University, who tested the feasibility of the computer application of these precision costing concepts in a Foxpro based algorithm as the basis for his 1992 Master's Thesis in Industrial and Systems Engineering.
>
> Miguel Serret-Alvarez, another Ohio University graduate student, who tested the application of these costing concepts to the design and costing of punch press tools as a part of his 1993 Master's Thesis in Industrial and Systems Engineering.
>
> Ricky Anderson, also my graduate student, who tested my concept of combining the SIMSCODER matrix part numbering system with basic

geometric group technology coding in his 1992 Master's Thesis in Industrial and Systems Engineering at Ohio University.

The results of these students' thesis research were of great help as reference materials in the preparation of this book.

The editorial assistance of Dr. S. Anne Sostrom, who reviewed the manuscript and offered many very helpful suggestions concerning its organization and readability, is also most appreciated.

E. Ralph Sims, Jr., P.E.

Contents

Preface	*v*
Acknowledgments	*xi*

1. The State of the Art in Manufacturing Costing — 1

- A. Introduction — 1
- B. Management's Dissatisfaction with Existing Techniques — 3
- C. The Accounting Impact of Capital Intensive Manufacturing — 4
- D. The Shrinking Direct Labor Content in Manufacturing — 5
- E. The Shrinking Direct Labor Content in Warehousing — 6
- F. The Management Implications of Automation and Computers — 7
- G. Some Suggestions that Have Not Been Followed and Why — 8
- H. A Different Approach in a High Tech Global Economy — 11

2. Generally Accepted Accounting Practice: A Review and Critique — 15

- A. Cash Method and Accrual Methods — 15
- B. Distribution Accounting — 17
- C. Direct Costing — 21
- D. Percentage Cost Accounting — 21
- E. Activity Based Costing (ABC) — 22
- F. Standard Costing — 23
- G. The Impact of Tax Law on Costs and Depreciation Schedules — 25
- H. The Product Budget/Estimating Process — 28

3.	**An Industrial Engineering Approach to Cost Accounting**	31
	A. A Different Philosophy of Management	31
	B. The Scientific Approach to Performance Measurement	34
	C. Work Measurement Methods and Standard Times	35
	D. Material Costs and Standards	39
	E. Equipment and Facility Costs	42
	F. Overhead Allocation Methods	44
	G. Building the Product Budget	46
4.	**Time Use of Facilities Product Costing**	49
	A. Freedom from Production Volume Variations	49
	B. Developing a Stable Cost Base	51
	C. Level-by-Level Cost Accumulation	53
	D. Cost Retrieval for Estimating and Design	56
	E. Precise Part/Operation Costing and Estimating	57
5.	**The Level-by-Level, Total Absorption, Standard Cost Concept**	59
	A. The Time Use of Facilities—"Use Rate" Approach	59
	B. The Total Absorption Standard Cost Approach	63
	C. Development and Application of the "Use Rate" Data Base	66
	D. The "Common Denominator" Concept of Material Management	69
	E. Level-by-Level Manufacturing/Warehousing Cost Application	70
	F. Level-by-Level Inventory Valuation and Product Costing	71
6.	**The Elements of Cost in Manufacturing and Warehousing**	73
	A. Labor	73
	B. Materials	78
	C. Indirect Operating Expenses	82
	D. Space and Equipment Cost	84
	E. General and Administrative Expense (G & A)	86
	F. The Cost Collection and Build-Up Procedure	88
	G. The Operating and Product Cost Statement	89
	H. The Cost Structure	90
	I. Summary of the Cost Philosophy	90
7.	**The Information Structure in Manufacturing and Design**	93
	A. The Integrated Information System	93
	B. The Flow of Information	96
	C. The Manufacturing/Numerical Bill of Materials	104

Contents xv

	D. The Structured Information Management System for Coded Operating Data and Engineering Retrieval (SIMSCODER)	108
	E. Level-by-Level Material and Operations Control	117
	F. The Matrix Type Information and Part Number Coding Structure	121
	G. Configuration Control and Documentation Coding	125
	H. Administrative and Operational Coding in the Matrix System	127
8.	**Material Cost Management and Control**	**129**
	A. Material Standards for Procurement	129
	B. Quality, Shrinkage, Chips, Drop Off, and Scrap Cost	133
	C. "Nest and Gain" Factor—The Cost of Schedules	136
	D. Inventory and Materials Management	140
	E. Work-in-Process Inventory and Just-in-Time Operations	148
	F. Inventory Costing and Valuation	151
	G. Getting the Material Cost into the Product Budget	152
9.	**Labor Cost**	**155**
	A. Organizational Payroll Philosophy	155
	B. Incentive Compensation Plans	156
	C. Measured Day Work and Salaries	157
	D. Fringe Benefits	158
	E. Labor Standards and Variances	159
	F. Machine and System Paced Operations	161
	G. Labor Reporting and Data Collection	162
	H. Defining Labor Cost Standards	163
	I. Getting Labor Cost into the Product Budget	166
10.	**Operating Equipment Cost**	**167**
	A. Capital Equipment Investment and Book Value	167
	B. Depreciation Schedules and Tax Law	168
	C. Maintenance, Modernization, and Replacement Costs	169
	D. Operational Support Systems and Utilities Costs	173
	E. Tooling and Operational Supplies Costs	173
	F. Utilization Factors and Idle Time Costs	174
	G. Equipment Utilization Cost Standards—"Use Rates"	177
	H. Getting the Equipment "Use Rate" into the Product Budget	178
11.	**Manufacturing/Warehousing Operating Burden or Overhead**	**181**
	A. Manufacturing/Warehousing Management Expense	181
	B. Indirect Manufacturing/Warehousing Labor	183

	C. Factory/Warehouse Overhead	184
	D. Non-production Materials and Supplies	186
	E. Contingent Manufacturing/Warehousing Materials	186
	F. Insurance, Interest, and Taxes	187
	G. Burden/Overhead Cost Standards	187
	H. Getting Operating Burden into the Product Budget	188
12.	**Facility/Building Burden or Overhead**	**191**
	A. Facility Capital Investment versus Book Value	191
	B. Land Investment and Non-depreciable Assets	192
	C. Facility/Housing Expense, Depreciation or Rent	192
	D. Utilities, Maintenance, Sanitation, and Security	194
	E. Facility Utilization Cost Standards—"Use Rates"	195
13.	**General and Administrative Expense**	**197**
	A. Definition of General and Administrative Expense	197
	B. Above or Below the Line Expenses	199
	C. Rationale for Including G & A in the Inventory Valuation	200
	D. Allocation of General and Administrative Expense	202
	E. G & A Cost Application Standards	203
	F. Getting the G & A into the Product Budget	203
14.	**Precision Costing of Manufacturing Operations**	**205**
	A. The Construction of the "Use Rate"	205
	B. Applying the "Use Rate" in Inventory Accounting	208
	C. Product Costing is Independent of Production Volume	208
	D. Precision Costing by Part, Operation, Department, Employee, and Product Level of Completion	209
	E. Group Technology and Process Sequence Costs	214
	F. Precise Materials Management in a Just-in-Time Environment	216
	G. Product Design Decisions and Predictable Cost Data	217
	H. Marketing and Pricing Decisions with Predictable Cost Data	219
	I. Better Responsibility and Activity Accounting	219
15.	**Precision Costing of Warehousing and Materials Handling Operations**	**223**
	A. Applying the "Use Rate" to Storage Space	223
	B. Applying the "Use Rate" to Materials Handling Equipment	229
	C. Application of Storage and Handling Cost to the Product's Inventory Value	233
	D. Distribution Center/Warehouse Costing	234

16.	**Sample Product Cost Calculations**	**245**
	A. The Product and the Bill of Materials	245
	B. The Make or Buy Decision	248
	C. Castings and Foundry Operations	249
	D. Computing Materials Costs	250
	E. Costing Purchased Parts and Components	256
	F. Computing Labor Rates and Costs	259
	G. Computing Work Station "Use Rates"	261
	H. Process Definition for "Make" Parts	265
	I. Computing Manufacturing Operations' Costs	272
	J. Computing Assembly Operations' Costs	276
	K. Computing Warehousing/Storage Costs	279
	L. Accumulating Level-by-Level Inventory Valuations	282
	M. Defining the "Totally Absorbed Product Cost"	286
17.	**Operations Analysis and Cost Variances**	**289**
	A. Cost Variances—A Management Analysis and Cost Control Tool	289
	B. Over- and Under-Utilization of Facilities	292
	C. Dealing with Labor Cost Variances	293
	D. Equipment Selection and Utilization Variances	295
	E. Dealing with Process Variances	297
	F. Dealing with Material Cost Variances	299
	G. Product Cost Variances and Their Causes	299
	H. Warehousing Cost Variances and Their Causes	300
	I. Information Feedback for Cost Control and Product Design	302
18.	**A Look into the Future**	**305**
	A. The Changing Industrial Society and Ecosystem	305
	B. Costing Computer Integrated Manufacturing (CIM)	309
	C. Capital Intensive Industry Requires Precise Costs	311
	D. High Tech Products Require Precision Costs	312
	E. Global Operations Require Precise Costing	313
	F. The Old Ways Are Inadequate—New Precision Costing is Required	315

Discussion Questions   **319**

References   **325**

Index   *329*

1
The State of the Art in Manufacturing Costing

A. INTRODUCTION

The nineteenth and twentieth centuries have seen the development and rapid evolution of both manufacturing technology and the related accounting methods which track and control its operations. In the late nineteenth century and the first quarter of the twentieth century, manufacturing systems progressed from craft and unit production to mass production. At the same time, management systems changed from groups of craftsmen working independently or in teams to mass production with moving assembly lines, specialized tasks, and division of labor.

During the second half of the twentieth century, the continuing industrial revolution evolved from a human labor and skill dominated stage through a heavy manufacturing and materials handling mechanization era, and into today's capital intensive and computer driven environment. Modern manufacturing operations utilize information technology, computerized management systems, computerized design (CAD/CAM), flexible manufacturing systems (FMS), robotics, Just-in-Time (JIT), and total quality management (TQM) methods. As we approach the twenty-first century, computer integrated manufacturing (CIM) is also gaining wider application in the industrialized world. Several trends are obvious in this evolutionary progression.

At each step, the capital investment per direct (and indirect) manufacturing employee has expanded so that direct labor cost is now a very small share of the cost of manufacturing and distribution.

During this same period, the emphasis of accounting procedures has shifted from maintaining simple cash records to providing more detailed accrual based corporate management systems. The introduction of income tax regulations and securities and exchange rules has superimposed government controls on the industrial and business worlds. Financial management and tax accounting disciplines rose in response.

The middle third of the twentieth century also experienced the expanded use of mass production methods and saw the development of new "scientific management" techniques. These new techniques attempted to relate labor and operations' time measurements and work schedule controls to financial and cost controls. The incursion of government regulations into management accounting practices continued.

The introduction of computers and automated production machinery in the last half of the century brought the means for real time precision accounting, but tax rules still dominated accounting philosophy and practice. Late in the twentieth century, industrial organizations began to change in response to the need for new competitive techniques and a "lean production" philosophy. This raised the requirement for more precise cost accounting practices.

"Generally accepted accounting principles" are the child of this evolution. "Generally accepted accounting *practice*" is based on traditional practices, the teachings of the university business schools, the information requirements of current business managers and their individual styles of management, and the past and current regulations of the Internal Revenue Service (IRS) and the Securities and Exchange Commission (SEC).

Accounting remains the feedback element of the management information structure. It collects, analyzes, and reports the financial behavior of the enterprise. The accountant's "statements" or reports are essential management tools which either reassure executives of the business's proper performance or trigger corrective action. If the accounting system collects accurate information but produces erroneous interpretations, management decisions are then faulty. Interpretations can be honest, accurate, and thorough, but their format, structure, or content can be based on a fundamental misconception of the nature of manufacturing costs, the need for the information, its use, and the level of detail required.

Today, most accounting systems continue to use percentages, averages, and ratios as the basis for evaluating performance and making business decisions. Most continue to distribute indirect and periodic expenses on the basis of varying production volume. Accountants still try to accommodate IRS rules when structuring their charts of accounts and in making analyses and reports. All these practices tend to distort the true costs of manufacturing operations.

Many manufacturing cost accounting systems lack a standardized human-machine legible information structure which includes a level-by-level manufacturing bill of materials and a matrix type data and information format. Because of this weakness, few existing systems fully utilize all the capabilities of data base software and random access computer technology to isolate and analyze the elements of costs in manufacturing enterprises. Few management information systems allow detailed retrieval of both cost and design engineering information while retaining the ability to allocate and define the fully absorbed cost of each operation for each part at every step in the process. These weaknesses limit the ability of the engineer to design to cost and the ability of the estimator to predict manufacturing cost.

Managers, academia, and the trade press are all seeking new approaches which provide a more valid and accurate definition of manufacturing costs and a sound basis for product cost engineering and production estimating. Many articles have been written which criticize past and present methods but none have offered a practical and convincing alternative to present methods. Many writers have pleaded with management to apply a new way of thinking to their decision making process, but none have led the way to a new logic. The trade press has also been vocal, but not persuasive enough in its prodding. The tools are here and they can be readily applied if management will change some of its thinking.

Because this state-of-the-art situation needs to change, one must examine the purposes of accounting and cost accounting. This requires reexamination of the accepted paradigms of management and the concepts that govern cost accounting today.

B. MANAGEMENT'S DISSATISFACTION WITH EXISTING TECHNIQUES

Management's dissatisfaction with the state of the art is an old story. Since the advent of mechanized random access accounting methods, dating back to the earliest disk file computers and magnetic ledger card accounting machines, managers have sought cost accounting methods which use data base technologies to more effectively and precisely identify and manage manufacturing costs. The isolation of labor and material costs through the use of job orders, time clocks, time keepers, and requisition procedures has been common practice for much of this century. However, managers and accountants continue to have difficulty with correct and equitable allocation of manufacturing burden, indirect production expense, and general and administrative expense.

Most accounting systems allocate these indirect costs on the basis of production volume or as a factor relating them to labor or material costs. These methods have proven to be inadequate in modern capital intensive, and often automated, manufacturing operations.

In addition to the need for better cost computation methods, it is essential to include cost calculations in every step in the design of the product. This is particularly important in complex product lines where common components are used in several products and partially completed parts are modified for a variety of uses in different products or different parts of the same product (i.e., lefts and rights). Precision costing of operations and parts is a key element of the design and manufacturing engineering process. This is particularly true when attempting to apply "concurrent" and "design to cost" engineering.

To achieve the required level of precision costing, it is necessary to blend the work measurement skills and techniques of the industrial engineer with the data collection capability of the computer system programmer, and the cost definition skills of the accountant. To create this mix and make the system work, it is necessary to overthrow old concepts of cost accounting and apply an engineering analysis approach to the cost collection procedure.

In this book, we intend to build a practical approach which is based on the current and future state of the art in computers, communications, and manufacturing technology. To achieve this goal, several factors must be considered.

C. THE ACCOUNTING IMPACT OF CAPITAL INTENSIVE MANUFACTURING

One of the most obvious results of modern manufacturing technology is the increasing level of the equipment and machinery investment required to produce a product. Automated metalworking machinery, plastic molding and extruding machines, robots, laser cutting and welding equipment, and the automation of food processing and packaging, glass making, and textile production have all increased the ratio of machinery investment to building cost, material cost, and labor cost. This has accentuated the importance of equipment burden as an element of production cost. It has also broken the "lock" between labor and machinery. Many modern processes and machines operate with little or no labor support, or intermittent worker participation.

In addition, the pressure of government, through environmental regulations and building codes, has had a heavy impact on building design and construction costs, resulting in an increase in the ratio of facility cost to material and labor expense. These costs must all be absorbed into the value of the product and recovered through inventory sales.

Many of the new technologies incorporate flexible manufacturing systems which can deal with short runs of a variety of product designs and frequent schedule changes through software reprogramming. This makes it necessary to charge machinery usage to a number of products during a single work period.

The State of the Art

Conversely, high-volume automation and process type manufacturing systems favor product specific "hard automation" which is less flexible and less responsive to product and schedule change, and often runs at a fixed speed. This allows the accountant to use a simpler cost allocation procedure in hard automation situations. In all cases, however, the high capital investment value of the machinery, compared to the material and labor used in the process, forces management's attention to indirect costs and requires the precise allocation of equipment burden and manufacturing overhead to the cost of the product.

D. THE SHRINKING DIRECT LABOR CONTENT IN MANUFACTURING

In the early days of the Industrial Revolution, the invention of machinery and interchangeable parts began to reduce the proportion of production cost which was generated by labor. The development of moving production lines cut the labor content in assembly operations and increased the capital investment component. However, every machine still required an operator and all assembly operations were manual. The equipment expanded the productivity of the worker and the market absorbed the output. The ratio of labor cost to materials and factory burden or overhead soon stabilized at approximately one third for each segment of manufacturing cost in the piece parts industries. This ratio still seems to be common in low technology American industries.

As automation entered the industrial scene, the productivity of the factory worker and the productivity of the administrative personnel who supported the manufacturing system accelerated rapidly. The early automation of stamping presses and lathes (screw machines) was accompanied by mechanization of materials handling operations (forklift trucks and conveyors) to meet production demands and cope with rising labor rates. This increase in capital investment, coupled with rising labor costs and more sophisticated materials, tended to maintain the one third, one third, one third pattern of labor, materials, and factory burden.

More recently, introduction of the computer into management systems and the computer control of production machinery has had a dramatic impact on this ratio. The first applications of computers in manufacturing management were in production and inventory control and payroll accounting. Initially, the computer reduced clerical and administrative staff at the price of a high investment in early mainframe systems. This also changed the ratio of factory office overhead to factory administrative labor. Where effective programs were in use, computer control also reduced work-in-process (WIP) inventory investment.

Computers also changed the office burden cost's relationship to the factory floor expense. These early systems provided better manufacturing control at a higher, computer based, total administrative cost.

The fallout from these early administrative technology developments was evidenced by the reduction of clerical and middle management staffing and the later widespread use of desktop computers, networks, and computer terminals. Evolution to a more competitive computer hardware market later resulted in an easing of the capital intensity of computer based manufacturing administration.

The computer also invaded the factory floor with computer numerical control machinery and later computer managed machining cells and flexible manufacturing systems. In addition to reducing the ratio of labor content in manufacturing, these computer based changes also reduced the linearity of labor time in relation to production volume.

Computers also reduced the decision making lead time in the marketing and manufacturing process. This has permitted the reduction of work-in-process and finished goods inventories. Although frequent and closely scheduled receipts and minimum WIP inventory are not new ideas in the automotive, appliance, food packing, and bottling industries, this computer based "real time" information capability made application of Just-in-Time management practices more practical.

In the factory and warehouse, we saw the development of computer numerical control machinery, automated guided vehicles (AGV), automated storage/retrieval machines (AS/RS), robots, and programmable controllers. These were coupled with such data capturing tools as automatic identification (bar codes, radio [RF], infrared, magnetic, optical, etc.) and computer tracking of machine operations and AGVS. This combination has contributed to the reduction of labor in the production process and in materials handling while also increasing the relative size of the capital equipment investment.

At the same time, the expanded use of technological solutions to manufacturing management problems has increased the need for engineers, computer technologists, and skilled maintenance labor. Since most of these workers are generally considered "indirect" because they do not actually change, handle, or touch the product, they also increase the ratio of manufacturing overhead or burden to manufacturing labor and material. In the late twentieth century, the labor share of manufacturing cost in many modernized operations fell from about 33% to a level of 5–15% (Rafish, 1991).

E. THE SHRINKING DIRECT LABOR CONTENT IN WAREHOUSING

The same trend has been evident in the distribution and warehousing function. The computer has replaced paper systems in the processing of order information and inventory records. This has accelerated order response and reduced the need for inventory to cover information lead time. The use of automated warehouse systems, which include automated storage/retrieval systems, automated

The State of the Art

order picking systems, automated guided vehicles, and conveyor systems, has replaced human labor in materials handling and warehousing operations. Larger highway and rail vehicles and freight containerization have reduced the labor content in transportation operations. All of these changes have increased the capital intensity of materials handling, warehousing, and transportation operations in industrial companies, while reducing the labor cost share of operating expense.

The heavy use of human labor in distribution operations is declining rapidly and the need for more precise information systems is increasing. In addition, because the Just-in-Time philosophy is receiving widespread acceptance, the order sizes have decreased and the transaction rates have increased. This has driven up the distribution share of the "landed cost of sales." In many industries, the distribution cost is increasing in relation to the manufacturing cost.

In many businesses, the physical distribution and materials flow/management functions have been isolated into separate cost or profit centers. They therefore require their own independent management and cost control systems. This is a recent departure from the former practice of including physical distribution and warehousing expense in the manufacturing overhead, with transportation often treated as a part of sales expense.

F. THE MANAGEMENT IMPLICATIONS OF AUTOMATION AND COMPUTERS

In general, the impact of computers and automation in modern manufacturing and distribution enterprises has been favorable. Usually, computers have resulted in more unified management control, better engineering and product design, easier information retrieval and manipulation, tighter quality standards, improved control of operations, more complete and accurate analysis of information for management decision making, and nearly "real time" data for responsive overall control of the enterprise.

Modern computer based automation has generally overcome the product design and market response limitations which have often been imposed by "hard automation" in the past. Computers have also provided the technical tools to permit modern cost accounting and operations management. But, as in all technical developments, there are some negatives to be considered.

It is essential to recognize that the computer is a "high speed idiot" and that it can only do what it is programmed to do. It is programmed by people who are "systems" knowledgeable. These computer technologists usually have a great depth of knowledge in the application of the computer to definable business and technical problems. However, many information system designers and programmers are not fully informed on the policies and decision criteria of the management culture in which their programs must perform. They may not (and often do not) have the breadth of business experience and the competent understanding to assure inclusion of management judgment criteria in their computer decision trees. As a result, they may design computer systems which erroneously discipline management behavior and limit the flexibility and responsiveness of management judgment.

It is also possible, as has been demonstrated by the Japanese (Toyota) KanBan system, to develop shop floor material and production control systems which operate without either paper or computers. These methods are an integral part of many Japanese Just-in-Time systems. They have a tendency to flag quality and performance errors, and reduce work-in-process inventories. These techniques also tend to increase the number of handling transactions and reduce the ability to capture detailed cost data for each operation and/or movement through the process.

When using a paperless and non-computerized system, detailed element costs must still be captured in order to evaluate the impact of "continuous improvement" programs and product design or methods changes. The application of a detailed standard cost system is one method for defining these costs.

It is, therefore, essential for management to carefully define the philosophical and policy structure of the corporate culture and establish the criteria for decisions before they are locked into the discipline of either a computer based management information system or a paperless Just-in-Time procedure.

The philosophical concept of manufacturing cost accounting is one of the areas which is most vulnerable to erroneous system structuring and overenforcement of computer discipline. In this volume we will attempt to define a costing philosophy which can be both management effective and computer compatible.

G. SOME SUGGESTIONS THAT HAVE NOT BEEN FOLLOWED AND WHY

There have been numerous attempts to develop a sound and acceptable basis for cost accounting in the modern computer based manufacturing environment. Computer Aided Manufacturing International (CAM-I) formed a coalition of professional accounting firms, private agencies, governments, and universities to exchange and share principles, ideas, and axioms in an effort to establish more realistic techniques for cost management and develop new cost accounting ideas. The Cost Accounting Standards Board also addressed many of the direct distribution costing problems. They attempted to make costing patterns more representative. They developed various standards to deal with problems such as the allocation of period costs to products and the separation of overhead costs into smaller pools to establish more meaningful bases (Sourwine, 1989).

Activity Based Costing (ABC) has also been reviewed and refined by many accounting professionals; among them Gilligan and Peavey (1990). "Activity accounting" was among the many ideas which CAM-I considered. Activity based costing (ABC) is one of the latest attempts to solve the cost accounting precision problem. Activity based costing assigns "cost drivers" to the various sources or elements of manufacturing expense. This approach allows a manufacturer to control costs by tracking those activities which are the sources of product costs (Brimson, 1988). This technique attempts to trace costs to the production basis of the activity consumed or performed on the product during the manufacturing process. According to Ostrenga (1990), the technique has two major attributes—activity based process costing and activity based product costing. Activity based process costing assigns costs based on the series of activities or drivers which are involved in the process. It also attempts to separate "value adding" expense from "non-value adding" activities. Activity based product costing assigns costs to the product by developing cost pools that vary with a common activity. The assignment of burden costs is based on allocation to the activity or cost driver as opposed to labor or machine hours.

However, activity based costing still treats general and administrative expense as a "below the line" cost and is production volume responsive with the resulting inverse manufacturing cost-to-production volume relationship. In some cases, the "non-value adding" costs are included in the inventory carrying costs. In the case of a minimum inventory system as in JIT, this could result in a distortion of the inventory (WIP) holding cost.

Machine Labor Costing was suggested by Schwarzback and Vangermeersch (1983) as a means to separate the cost of the key machines from other cost sources. The machine costs are broken into variable and fixed components, and the standard and actual costs are calculated separately for each machine. Machine costs are charged to the product on the basis of machine labor hours. This method provides good control of manufacturing overhead and stresses machine utilization. It also defines the differences in machine hour costs based on the type of machine and allows the comparison of operations based on the choice of machine and process. However, it also uses departmental overhead allocation and does not recognize general administrative and work-in-process costs.

Process Costing (Woods, 1989) assigns costs to processes by accounting for the degree to which the processes cause costs to be incurred. This method requires the cost of supporting departments to be allocated to specific processes based on a measure of the service used and the costs incurred in production. Material cost is assigned in the normal manner. This method focuses on the cost of the process instead of direct labor and on process instead of product. However, it uses the variable bases of labor and machine time. It can provide good supervisory control of operations, but it does not separate overhead costs within each department and, therefore, cannot provide a valid product cost or inventory valuation.

The *Productive Hour Rate Costing* method which was suggested by Ostwald (1984) calculates the productive hourly rate by adding the machine hourly rate and the direct labor hourly rate. In this concept, all of the indirect costs are included in the machine hourly rate. However, it does not address non-machine operations or the true impact of costs that are not linearly related to machine hours. It does closely approach a total absorption cost system for machining operations.

Technology Accounting is a philosophy introduced by Brimson (1988 and 1989). This technique dictates that technology costs should be traced directly to the product that is benefiting from them. According to Peavey (1990), this technique causes amortization of idle machinery, which increases apparent overhead costs when there is little production and inflates the production costs. This encourages constant production in order to maintain a target cost per unit. A suggested adjustment to this method would use production based amortization of the machinery, but this would also result in a variable base for handling a fixed cost and would ignore obsolescence factors.

Life Cycle Accounting is "the accumulation of costs for activities that occur over the entire life cycle of a product from inception of a product line to abandonment by the manufacturer and consumer" (Peavey, 1990). This is a valid concept for use in the analysis of the marketing viability of a product and definition of the overall cost effectiveness of a product design. It is not a good tool for the management of product costs in manufacturing or for definition of inventory values for work-in-process or finished goods.

All of these suggested methods suffer from a common weakness. None of them isolates definition of the cost of the product from definition of the overall performance of the business. They all attempt to allocate all of the costs of the business to the product on the basis of the level of production and/or a variable production related base such as labor hours or machine hours. In the final analysis, each of these techniques causes the apparent cost of the product to increase when production volume decreases and to decrease when the output increases. This does not take into account the fact that production costs are a function of time use of the facility and that unused or overused capacity is usually a function of sales effectiveness, not manufacturing performance.

In a capital intensive industry this is particularly deceiving since the cost of idle or overutilized equipment is a major burden factor and it can distort apparent costs and result in erroneous product design and marketing decisions.

H. A DIFFERENT APPROACH IN A HIGH TECH GLOBAL ECONOMY

One of the cost features of a global economy is the wide variation in the cost of labor, materials, taxes, and overhead at different manufacturing sites. At the same time, another feature of a global manufacturing system is the relative uniformity of technology, methods, equipment, and products in the various markets of the world.

In spite of dollar-a-day labor in countries like India and Indonesia, modern machinery and methods are used there in the same way as they are used with $15 per hour labor in North America. Placement of a plant at the source of the raw material to reduce bulk transportation costs does not change the standard processing method or the capital equipment used to convert the materials. For example, the complexity, manufacturing process, and required street performance of a motorcycle is the same whether it is produced in Japan, India, Europe, or North America.

Another consideration is the basic manufacturing philosophy. In Japan, the management attitude states that "inventories are a drug with respect to production" and they cover up errors in scheduling and quality management. The Japanese also feel that "one hour of inactivity (in a work station or plant) is better than an hour of production that must be sent to a warehouse" or scrapped (Giorgio Merli, 1987/1990). By contrast, most Western manufacturing managers strive for efficiency through the optimum utilization of the capital facilities and accept WIP or finished inventory investment as a normal adjustment factor in the scheduling of continuous production flow. Although they cringe at the thought of idle labor or machinery "downtime," they are beginning to accept the Just-in-Time philosophy in materials management.

The continuing shift to more capital intensive manufacturing and distribution operations makes the need for a more precise and product oriented costing system quite apparent. Management must be able to compare product designs, manufacturing methods, and production performance on a valid, stable, and equitable basis. A new, volume independent approach to the problem of product and manufacturing costing is required. It must be based on such stable and global standards as time, space, and units of material.

The ability to precisely estimate the cost of products during and before their design is another primary objective of cost engineering. It is usually desirable to design products to cost targets in today's simultaneous or concurrent engineering environment. A realistic and fully absorbed standard or estimated cost for each operation and/or part is an essential element of cost based engineering.

It is essential to isolate the operation and product cost from the unused or overused overhead expense. Computation of the break-even point in projecting the expected production or sales volume is a separate issue from the calculation or estimation of the true cost of making the product.

The "Precision Engineered Costing System" is a method of costing used by both piece part and continuous process manufacturing operations with particular focus on the cost finding impact of capital intensive, high tech operations. It is a composite of generally accepted accounting practices which are reoriented to achieve a true, total absorption, time use of facility based, level-by-level cost allocation to work-in-process and finished goods inventories.

To provide a basis for this type of costing, it is necessary to develop a "use rate" or rental charge for the time use of the facilities. The time use of facility approach is implemented through the application of a "use rate" or rental approach to the complete absorption of all period and indirect expenses into the product based on work station and equipment operation or use time. This "use rate" is the vehicle for collection and application of *all* period, indirect, burden, overhead, and general and administrative costs into the product and operation on a level-by-level, operation-by-operation, time use of facility basis.

Unused facility expense and overhead should *not* be charged to the product. The production schedule or volume based idle facility costs (or overabsorbed overhead) should be treated as a corporate overhead expense or profit and be charged to the profit and loss statement. This cost is a measure of overall corporate performance and marketing success. It is not usually a result of manufacturing operations. The "use rate" should be adjusted on a periodic basis (six months or one year) to accommodate and absorb variance trends and the impact of inflation.

This type of costing system requires a level-by-level approach to manufacturing (and engineering) management. By definition, level-by-level material control is "production control." In addition, it provides the basis for cumulative, level-by-level, cost allocation to the product as it flows through the process.

This approach can be applied by combining matrix type numerical bills of materials and sortable part numbers with the total absorption standard costing of each operation, part, and product. Time is the only positive and global basis for the measurement or definition of any activity. These standard costs should be built on the stable base of time use of facilities and labor, plus the standard cost of materials used in production.

This system divorces production volume based cost variations and tax based accounting practices from the process to develop a true and more complete measure of manufacturing costs. Each operation assumes its full share of all labor, material, burden, overhead, and general and administrative expense. This results in a level-by-level cost structure which is built in a bottom-up manner based on the time use of the manufacturing or warehousing facilities.

The State of the Art

The system provides for a true activity based cost system and realistic responsibility accounting. Idle capacity costs and overabsorbed burden and overhead are allocated to the profit and loss accounts and not to the product. Unused capacity or above plan production can be properly charged, or credited, to the success or failure of marketing or product development and not to the manufacturing operation.

To apply this costing concept, it is necessary to develop a matrix type numerical manufacturing bill of materials system and sortable human-machine legible part, product, and operation numbering systems. These systems will provide for the "tagging" of cost, production, and engineering data. The coding system will also aid in engineering information retrieval, cost accumulation, inventory valuation and analysis, operations data management, detail cost allocation, and parts standardization.

To set the stage for some of the concepts presented in this book, and with the foregoing comments as a background, it is appropriate to summarize some principles which are parts of the author's philosophy of management. The following are some of the thoughts or concepts upon which of this book is based.

- "Price is independent of cost! Price is set by the marketplace and the competition."
- "Cost is a function of product design and management skill."
- "Profit is the difference between cost and price. It is the scoreboard of top management's performance."
- "Time is the only reliable, measurable, predictable, and incorruptible element in the cost equation. It should be the basis for defining and managing all other elements of the operation."
- "All costs should be allocated to the product on the basis of the time use of facilities and labor, plus material, at every operation level or step in the flow of materials through the enterprise."
- "The flow of materials and product through an enterprise is the physical manifestation of the flow of funds through the business."
- "The level-by-level control of the materials and products flowing through the process assures management's control of the enterprise."
- "All production costs must flow into the valuation of inventory (raw material, work-in-process, finished goods) at each process level."
- "All production revenues are derived from the disposal of inventory."
- "Overhead absorption variances which result from capacity utilization variances are profit variances and not product cost variances."

The "use rate," total absorption, time use of facility costing philosophy presented in this book attempts to apply these principles in order to achieve a more correct and accurate definition of manufacturing and product cost.

2
Generally Accepted Accounting Practice: A Review and Critique

A. CASH METHOD AND ACCRUAL METHODS

There are two basic approaches to business accounting, and, to some extent, the choice of the method to be used is dictated by tax law. Many small businesses and individual business people use the cash method of accounting. Recent tax law also requires professional practitioners and personal service businesses to use the cash method for tax purposes. Most manufacturing companies and distribution enterprises that maintain inventories and/or have accounts payable and accounts receivable records use the accrual method of accounting. A few firms use a combination of the two systems. The resulting profit and loss statements and balance sheets present very different pictures of the condition of the enterprise.

In either an accrual or cash system, it is essential to have a payroll and payroll tax account to comply with labor and tax laws and to record labor utilization. A manufacturing bill of materials (MBoM) is required to define the product, its assembly and manufacturing sequence, and the inventory levels in the production process (Figure 2.1). In addition, a four balance inventory record (on-order, on-hand, allocated, and available) is necessary in order to include the sequencing or scheduling factor in materials management, to manage work-in-process (WIP) and finished inventories, and to control manufacturing supplies and purchases. Although Just-in-Time (JIT) claims to avoid work-in-process documentation and to eliminate or significantly reduce WIP, the scheduling and costing of production still requires "inventory windows" (Figure 2.1) or

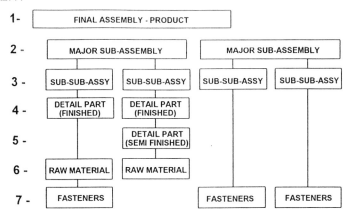

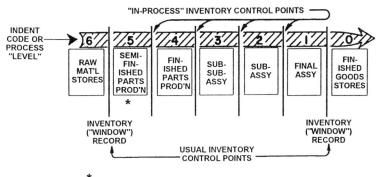

Figure 2.1 Process control point–indent level relationship.

control points between operations, at bill of materials "indent levels," and whenever the WIP item changes its identity, physical characteristics, or control status. It is also necessary to use "material requirements planning" (MRP) concepts in order to establish the required lead times when developing level-by-level production schedules.

In the cash accounting system, a check register or journal is sufficient to record actual income and expenses. The entries are then coded for allocation to tax categories and cost accounting (project, work order, or product) reports.

In addition, the accrual system requires double entry records for accounts receivable and accounts payable, capital equipment and/or real estate accounts, depreciation and maintenance charges, long term liabilities and debt service accounts, interest records, and a tax accrual account. In most systems these basic accounts are supported by detailed sub-accounts which are then organized by operational or materials category (office supplies, production materials, hourly and salary payroll, taxes, insurance, utilities, travel, etc.). Most manufacturing enterprises use the accrual method of accounting.

B. DISTRIBUTION ACCOUNTING

"Distribution" Accounting has been the most common technique used for cost accounting and product costing since the beginning of this century. This method accumulates all of the period, indirect, overhead, and operating burden costs incurred during a specific time period, adds them to the direct manufacturing costs, and balances the accumulated result against actual expenditures for the period to identfy variances in performance. The assumed indirect cost of production is often computed by dividing the total indirect expenses for the period by the amount of product produced in that cost center during that period. This figure is added to the direct labor and material costs to define the product cost.

Distributive cost accounting is based on the premise that all costs incurred within a period must be allocated to the product(s) produced (or sold) during that period in such a manner as to fully balance the actual expenditures with the assigned costs. However, this approach results in the variable assignment of indirect costs as production volume and utilization of facilities fluctuate. The apparent cost of the product will obviously vary inversely with production volume.

For example, if a factory or warehouse is operating at three-quarters of its capacity, the distribution method would assign the cost of carrying the idle 25% of the facility to the products manufactured or stored during that period. This would inflate the apparent unit cost of manufacture or storage, load the inventory value, and distort the cost-price relationships. Conversely, the facility which converts from one to two shift operations, or schedules overtime, would appear to have a reduced unit cost because of a broader time base for overhead absorption. However, in each case, the actual time use of the facility or cost per cubic foot of warehouse space per hour or day for the manufacture or storage of the product would be the same. Only the accounting relationships would change.

A refinement of the distribution accounting technique can recognize the need to isolate the cost of idle or overutilized capacity and still use the production volume approach to costing the product being manufactured. Figure 2.2 shows the origin of the "use rate" concept based on conventional, production based, distribution accounting concepts.

In this case the computed "use rate" is a gross figure based on the total burden and overhead. It is applied on the basis of production volume. In many

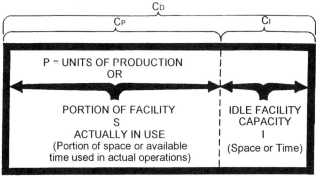

PRODUCTION OR WAREHOUSE FACILITY

C_D = Total burden and overhead for whole facility

C_P = Total burden and overhead for portion of facility in actual use (area or time share)

C_I = Total burden and overhead for idle portion of facillity (space or capacity

① DISTRIBUTION (CONVENTIONAL) METHOD $Cu_D = \frac{C_D}{P}$
where Cu = unit burden and overhead allocation to production

② USE RATE METHOD $Cu_P = \frac{C_P}{S}$

③ IDLE CAPACITY COST AS A UNIT VARIANCE TO P & L $Cu_I = \frac{C_I}{I}$
where I is in potential additional unused capacity in space units or time

Cu_I should be charged as an unabsorbed variance to profit, sales or management expense.

Cu_I should not be charged to product or manufacturing expense.

Figure 2.2 Applying the "use rate" concept in a conventional distribution accounting system.

cases the indirect and period costs are applied as a ratio or multiple of direct labor or material cost. These methods result in a volume based product cost estimate. Although this is a step in the direction of improved precision, it still produces a variable and incorrect volume based cost figure.

Distribution accounting techniques result in a moving basis for analysis of performance. As a result, the conditions for success or failure cannot be pre-

Generally Accepted Accounting Practice

cisely defined. The bottom line profit performance is definable but the internal cause and effect relationships which the profit performance reflects are obscured by the relativity of the information input.

In such systems, all fixed, direct, and variable costs are absorbed by the work-in-process and finished goods inventory and, in turn, become part of the inventory valuation. In most such accounting systems, the general and administrative (G&A) cost is ignored at the factory level and treated as a "below the line" cost. It is not usually included in the valuation of the inventory. The product's unit cost is derived by dividing the accumulated direct and indirect costs by the number of units produced in the specified time period.

A primary drawback of the distribution type costing method is that the volume of production and the utilization of the facility fluctuate. This results in an inconsistent assignment of indirect costs and overhead to the computed product cost. (Sims, 1968; Emig and Mazeffa, 1990). For example when a facility converts from one shift to three shifts, it will appear to have lowered the product unit cost. In reality, the product has consumed the same time use of facility, labor, and material.

However, because the fixed and indirect charges are redistributed over more hours, the computed cost of the product is altered. It is therefore desirable to recompute the standard cost to be sure that the impact of the schedule change is recognized. The cost of time in the "use rate" of the time use of facility will be based on a new number of scheduled operating hours.

This is the means by which the time use of facilities system adjusts to production volume's scheduling impact on computed costs. It should only be used when a permanent or long term change is made in the scheduled hours of operation. Short-term changes and irregular schedules should be treated as variances.

If this variability problem is not addressed, management may make erroneous pricing and design decisions based on faulty scheduling and cost information.

When using distribution type costing techniques, if utilization of capacity is reduced, the apparent cost of the product increases in the face of reduced sales. In such a case, the manufacturing executive may be faulted when the problem is a lack of sales. The apparent rise in cost may also trigger an inappropriate price increase.

Another weakness often found when these methods are used is in the allocation of overhead and burden calculated by using a wage cost percentage or a unit charge based on direct labor hours or machine labor hours (Figure 2.3). These are variable bases which fluctuate as the volume changes. In capital intensive industries, direct labor has little to do with the accrual of overhead cost because the labor/machine relationships are non-linear. Modern skilled labor often simultaneously operates multiple machines and/or relates to the machin-

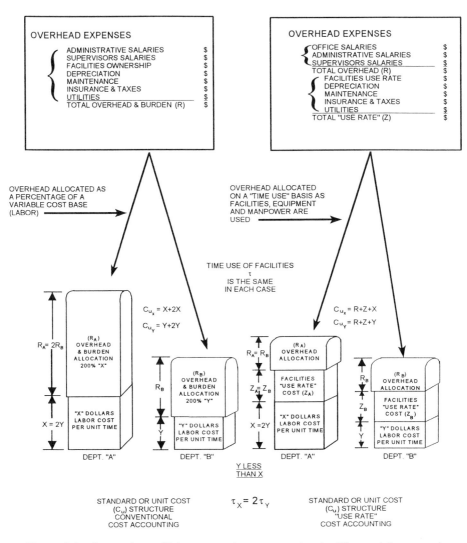

Figure 2.3 Comparison of labor percentage versus standard "use rate" approach to assignment of overhead and facilities burden cost to product cost.

ery on an intermittent schedule. This makes it difficult to assign costs with a high degree of certainty (Sutton, 1991; Dhavale, 1988; Sourwine, 1989; Brimson, 1988; Gilligan, 1990).

Emig and Mazeffa (1990) stated that the distribution method also fails to recognize that product diversity drastically increases or distorts apparent overhead costs. They suggested that in a distribution accounting system, indirect costs, or costs which are usually averaged across all of the firm's products, need to be identified and traced to the specific parts or products that generated the overhead cost.

Brimson (1988) recognized the fact that distribution accounting does not penalize overproduction. In his study, it was hinted that the idea of overhead absorption based on production volume encourages companies to keep utilization in excess of demand since it lowers the apparent cost of the product. This is the concept of "efficiency" based on full utilization of capacity which has been overthrown by the Japanese concept of continuous improvement and production based on quality and the acceptance of downtime in preference to WIP inventory build up (Merli, 1990). In addition, the distribution method does not isolate the costs of unnecessary activities or unused capacities.

Ashtan and Holmlund (1991) stated that because of the absorption effects, the reduction of work-in-process and finished goods inventories gives management a false status impression by reporting a decrease in profits even though the actual cash flow increases.

C. DIRECT COSTING

Direct Costing is a variation of distribution type costing. In direct costing, the fixed or period elements of overhead are separated from variable overhead costs and are treated as a period expense. A weakness in this system is the tendency for inventory valuation to be lowered because it smooths the effects of period costs by assigning them below the line.

The objective here is to provide a basis for incremental cost and profit analysis. The theory states that if only the variable direct and indirect costs are assigned to the product, the varying relationship between volume and capacity or volume and profit can be more easily defined. From the point of view of the top executive or marketing manager, this technique has the advantage of providing a better picture of the relative effect of each product on marginal profit performance. It also tends to lower the apparent inventory valuation because it smooths the effect of period costs. The conventional distribution technique picks up these costs on a production date basis and carries them into the inventory. Direct costing assigns them to a below the line status. This method, although a bit more definitive than the distributive procedure, still lacks the ability to precisely define and recreate the operating conditions which produced the success or failure to meet planned performance criteria.

D. PERCENTAGE COST ACCOUNTING

Before the introduction of computers, accountants and managers developed the practice of using percentages as the basis for distributing indirect expense to the cost of a part or product. In most pre-computer manufacturing operations, the ratios of labor to material and overhead averaged about 33% for each segment of cost. With this traditional distribution, it was logical to allocate overhead and indirect expense to the product as a percentage of either labor or materials.

The problem with this practice is the variability of both the labor rate and the material cost from operation to operation. In secondary and assembly operations, there is generally no new material to be added. Therefore, the most predictable cost element (material) is only useful as a basis for costing in the initial operation of a process.

Labor rates also vary with the person assigned to the task and with the schedule (shift work and overtime) of the job. Both material and labor costs also vary with time, as do many of the elements of the overhead cost. Therefore, any assignment of a percentage of the value of labor or material as the basis for allocation of overhead or burden faces the problem of a variable base. In many cases this variation is modified by setting a period standard labor cost or material cost and allocating a standard overhead or burden value to the labor hour or material unit. This provides a uniform standard cost. However, if burden and overhead calculations are based on a distribution type cost system, it will not produce an accurate product cost.

All of these systems—conventional distribution costing, direct costing, and percentage costing—attempt to calculate product cost in a more representative way to reflect true cost performance. However, they all suffer from the same errors because they distribute indirect and overhead costs to the product on the basis of production volume and/or labor utilization.

E. ACTIVITY BASED COSTING (ABC)

Activity based costing (ABC) attempts to identify costs with the activities which generate them. This technique uses the concept of "cost drivers" and attempts to separate value adding activity costs from non-value activity costs. The ABC approach also tries to address life cycle costing, technology costing, and financial accounting requirements. Activity based costing is a step in the right direction. However, it still relates indirect expenses to cost drivers and product cost on the basis of production volume. ABC also attempts to interlock financial management requirements with the costing of manufacturing operations. Some income tax accountants say that ABC costing of inventory is not acceptable for tax purposes because the authorized method for valuation of inventories is defined in Section 126 of the Tax Code. However, this criticism is inappropriate when using data base computer systems since inventory values can be computed differently for different audiences.

The proposed time use of facility and "use rate" approach presented in this book isolates product cost from production volume and financial management criteria. It is a refinement of the concept used by activity based costing. Precision engineered costing relates all indirect costs to the facility and its time use. The time use costs of the work station define its "use rate." The "use rate" becomes the cost driver in defining the time cost of the "activity" or operation. This separates the definition of manufacturing costs from dependence upon such variables as plant-wide production volume, labor utilization, and material costs.

Any of these methods will produce usable variance data if the calculations are correct and consistent with the actual cost data collection system. However, as a basis for computing or predicting true product costs, the standards which use a distribution technique for assigning overhead, burden, and G & A expenses will have the same error factors as the technique used to produce them.

F. STANDARD COSTING

Standard Costing is a broad term which generally implies that one of many methods has been used to define a base line cost against which actual costs can be compared. Standard cost systems can be based on any of the above described methods of calculating costs. The variance between the actual reported cost and the standard is used as a measure of performance deviations from plan and a guide for correction of problems. In most operations, the variances will be small and random if the standard is correctly calculated and the operation is running according to plan.

The concept of a standard cost system can be demonstrated by using a military analogy (Figure 2.4) of an artillery piece firing at a fixed target and working with a forward observer. In the gunnery case, the location or position

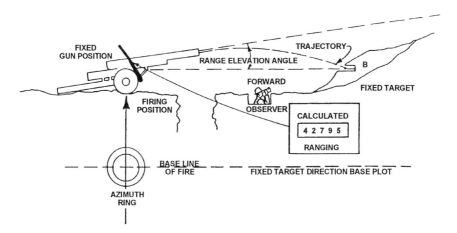

Figure 2.4 A military analogy of the standard cost concept.

of the gun, the azimuth and range to the target, and the computed trajectory of the round, are all defined by the gunner from transit sightings, map coordinates, and ammunition data. These are the "standards" upon which he bases the firing operation. However, the forward observer (or inspector) is capable of measuring the impact error or variance which may result from windage or errors in the siting and targeting of the weapon. The forward observer's variance data will provide correction information for the gunner to hit the target.

The development of engineered standard costs can provide a tool for both management and engineering estimating. An engineered standard cost is one which is based on a well-defined work task.

The use of either stopwatch time study data or predetermined time standards (MTM = Methods Time Measurement), and/or a machine paced operation, provides the basis for the standard. In the application of engineered standards, it is customary to establish the labor component in minutes per piece or unit of production for each operation and to define the standard material usage per piece. If the operation is machine paced, the labor content is usually internal to the machine cycle time. In most distribution based systems, the values of the labor and material are applied on a current or period standard cost basis and burden or overhead is applied as either a percentage factor or a separately computed standard cost per unit of time.

Precision costing is a refinement of the engineered standard costing technique which allows more complete analysis and control of all phases of the operation. In this procedure, all elements of burden, general and administrative expense, and indirect and direct labor costs are related to the time use of facilities, and all elements of direct labor and material cost are related to the detail part and operation. The cost element variances developed in precision costing are capable of pinpointing cause and effect relationships in all elements of cost and performance. The standards are built on a level-by-level basis which aids in automated or computer based design and estimating. The standards permit level-by-level valuation of raw, in-process, and finished inventory on an operation-by-operation basis. This technique also permits extraction of cost by machine or work center, department, worker, job order, product, and part.

It is obvious that standards will vary with time. However, if the variances during a standard period are random, this should indicate that the standard has been computed correctly. If the variances exhibit a trend which is either rising or falling, the standards should be adjusted to accept reality or the process must be corrected or altered to regain the standard performance. In most cases period standards are set for a year or for a production run. Figure 2.5 shows a method of dealing with a valid variance trend in the material component of a cost standard.

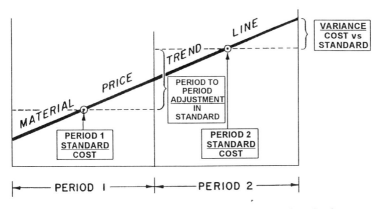

Figure 2.5 The concept of adjusting standards to absorb element cost trends.

G. THE IMPACT OF TAX LAW ON COSTS AND DEPRECIATION SCHEDULES

Income taxes, real estate taxes, inventory taxes, and personal property taxes are major "period cost" items which must be dealt with as a part of the non-productive or overhead cost of doing business. The amount of income tax paid by a firm is directly related to the gross profit earned, and is a "below the line" item which is not included in the manufacturing cost of the product. It is a factor in computing margins and prices. The product cost impact of the income tax law is found in the calculation of the basis for taxation. Other tax impacts are derived from taxes which are not earnings related.

All of these taxes vary with time, tax legislation, and the performance of the business. In addition, since tax laws usually specify certain accounting procedures or rules for calculating profit, inventory value, or other bases for taxation, the tax base and its cost impact is government driven.

Tax rules include, among other items, the allowable depreciation "life" of capital equipment and buildings, the cost basis for inventory valuation, and the treatment of the period cost of financing capital equipment, buildings, and inventory (interest). Federal tax law now mandates the use of the MACRS depreciation system and a depreciation period of seven years for most manufacturing facilities. These rules can change at the whim of Congress or state legislatures. In many cases the tax rules impose unrealistic values and service life times, and thereby affect the accuracy of cost calculations.

To improve cash flow and profits and to minimize tax cost, financial managers also try to use accelerated or variable depreciation schedules which charge off larger portions of the capital value of buildings and equipment in the early years of their investment life. These practices include such procedures as "double declining balance" and "sum of the years digits" depreciation methods in place of the more theoretically valid "straight line" calculation (Figure 2.6). These rapid write off practices have a multiple impact.

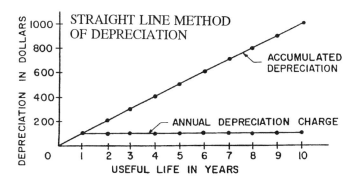

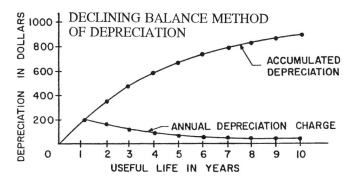

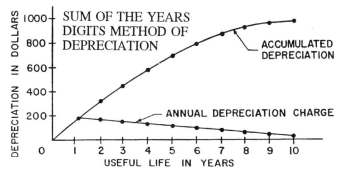

Figure 2.6 Comparison of the effect of different methods of depreciation.

Generally Accepted Accounting Practice

On one hand, in firms with rapidly progressing high technology, the tax life of an asset may be much longer than its functional life or obsolescence life. On the other hand, the productive life of conventional machinery and buildings is often longer than the allowable tax life of the asset. This situation has a double-edged effect.

If the asset is obsoleted and disposed of before its tax life, a straight line depreciation rate undercharges the overhead account. An accelerated rate could overcharge if compared to a sinking fund type of capital recovery program. In any case, the charge is an arbitrary one and often has little relationship to the true cost of owning the asset.

Although any solution to this dilemma is, by its very nature, an arbitrary one, the financial manager must conform to the mandated requirements of the tax laws. At the same time, the operations or manufacturing manager and general manager need a reliable and defensible method for defining the true cost of manufacturing the product. They need to realistically absorb all of the overhead and capital recovery costs into the product cost at each stage of its production. This need for precise and detailed cost data throughout the process requires level-by-level costing of inventory and total absorption of all costs.

This requirement leads to the development of the "use rate" concept of overhead absorption and inventory costing. The "use rate" uses a pragmatic and practical approach to capital investment recovery as an element of overhead cost. It is similar to a rental rate in that it captures all costs on a time use basis. This "use rate" approach to inventory costing will be discussed in detail later.

Income tax and depreciation schedules are not the only external forces imposing procedures on the accounting system. One variable "fixed" cost of operations is derived from real estate taxes. These tax rates can often be altered by legislative action, and periodically by changes in real estate assessments and appraisals. These costs must also be included in the overhead absorption procedure.

Another government sponsored expense is the personal property tax, which varies from one jurisdiction to another. For example, at the time of this writing, and subject to change by the legislature at any time, one state levied a tax of 4% on 70% of the book value of inventory on the thirty-first of March each year. However, the method for calculating the book value of the inventory varies from company to company, and over time, based on the choice of accounting methods. The choice between FIFO (first in-first out) and LIFO (last in-first out) valuations and the impact of the rates of inflation and/or deflation of the market and the economy alter the value of the inventory and the cost of the tax. This tax, then, becomes another variable period cost factor.

This same state also taxed "personal property," which is defined as equipment, furniture, and negotiable securities. In this case, the value of the items varies with their age and the method of depreciation in use. The value of securities varies with the market. This is another variable period cost factor.

For the purposes of this discussion, labor related taxes have been omitted from this section. Labor based taxes are normally considered to be a part of the fringe cost of labor and will be covered as a part of the consideration of labor expenses.

H. THE PRODUCT BUDGET/ESTIMATING PROCESS

All of these accounting decisions and procedures can produce varying types of skewed information which can affect the validity of the product budgeting and estimating process. The resulting cost computations can affect product design and process engineering decisions.

There are several common approaches to product estimating or budgeting. In many cases, rule of thumb estimates are used for engineered product proposals and even for pricing of the finished product. A good estimator can often be as accurate as the accounting system of the enterprise, and the final cost may be within 1or 2% of the estimate. On very large projects, techniques such as the critical path method (CPM) and the program evaluation and review technique (PERT) have been used to deal with the uncertainty of the program over time.

The CPM and PERT techniques are normally computerized. They can test a variety of data values to estimate the probable error range and to provide a basis for estimating and pricing. In big operations, it is also customary to admit the probability of error and to include a contingency factor. Modern computer technology and computer based forecasting methods use statistical theory in concert with financial data and simulation of the manufacturing operations to predict performance and estimate costs.

In any of these methods, the product budget or estimate is only as good as the basic data used in the process. Errors generated by distribution type accounting and conventional cost calculations are carried into the most precise and sophisticated computer based estimating and budgeting procedures.

The selected approach should be responsive to the price sensitivity of the market and the magnitude of the product cost. For example, in costing beer bottles, the estimator deals with a simple, single part product which is process manufactured on very expensive machinery. Beer bottles are made in long runs at the rate of hundreds per minute, and with their competitive margin computed in fractions of a penny.

Conversely, the manufacture of high tech machinery usually involves many high-priced complex parts and components, labor intensive assembly operations, and a more function/quality sensitivity and a less price sensitive market. The precision of the data in the bottle plant is very critical to price com-

Generally Accepted Accounting Practice

petitive estimating. In the machinery plant, the completeness of the data system is the most critical factor in the correct estimation of machinery costs. In both cases, erroneous data, derived from inadequate accounting procedures, can have a critical impact on the success of the enterprise.

All of these variables must be accommodated in the development of a precise manufacturing cost accounting system. Because of the complexity of the information relationships this manufacturing cost system must handle, it should be computer based. Modern computers can use data base type information systems with large data files and multiple access and report capabilities. It is therefore feasible to produce reports which are audience oriented and subject specific.

The computer can use the same raw data to provide a variety of focused reports. It can produce tax oriented financial information and statements, and financially oriented balance sheets and profit and loss statements. A computerized management information system can also produce function specific operating reports, and market oriented product performance reports. For manufacturing and engineering, the system can provide a stable and well-defined basis for measuring and predicting product cost and production performance.

Accurate and correctly focused management reports are essential to the decision making process. They provide the basis for good management planning and control. However, the reports must be based on reliable and precise cost information, accurate factory performance data, and mature statistical analyses.

3

An Industrial Engineering Approach to Cost Accounting

A. A DIFFERENT PHILOSOPHY OF MANAGEMENT

The foregoing chapters illustrate that, historically, much of manufacturing management's philosophy has been dominated by financial considerations and a top-down approach to data collection and analysis. The "distribution" approach to accounting and its responsiveness to tax law considerations has limited the development of accurate shop floor information related to manufacturing cost. Because the application of computers in industry began in the accounting area, this resulted in the establishment of in-place systems which focus on financial management goals. Later development of material control and manufacturing management systems was often hampered by these in-place and incompatible financial management systems. This made the development of detailed level-by-level manufacturing cost control procedures more difficult. The resulting financially oriented cost systems are now being subjected to criticism. There is a growing management desire for a new approach to manufacturing costing. The field of industrial engineering may provide an answer.

In contrast with this financially based, top down approach, the industrial engineering profession developed its philosophy and technology from the bottom up. In the early days of "scientific management," which led to the industrial engineering profession of today, Fredrick W. Taylor and Frank and Lillian Gilbreth approached task management and work measurement from the work element and time measurement point of view. They recognized that time is the only valid and incorruptible basis for performance measurement and they tied their definitions of work elements to the measurement of time. They developed cost data by relating labor time and wage rates to the definition of a task and the time it took to complete the work.

Although very simplistic and incomplete, their approach provided the basis for the analysis and costing of the work performed. It also initiated the bottom-up approach to cost definition and analysis. For many years, industrial engineers have used this approach in the comparative analysis of methods and operations, the definition of production costs, product cost estimating, and the design of wage incentive programs. These time and cost standards only tangentially related performance to quality and defects management. Because of generally accepted accounting practices still in use, indirect costs were applied to the product using production quantity based distribution, or as a ratio or percentage of labor or material costs. Both of these procedures resulted in the distortion of total product cost.

Work measurement is still a significant element of the industrial engineer's professional bag of tricks. The traditional roles of work measurement include the measurement of people's performance, the timing of machinery operations, the definition of direct versus indirect labor time, and the timing and measurement of space utilization. The newer philosophies of manufacturing management try to include the economics of quality and inventory ownership, along with the effects of quantity, as a part of cost effective Just-in-Time management.

Modern industrial engineers use new computer based mathematical analysis tools to accomplish their cost definition objectives in a more sophisticated and complete manner. Time measurement is still the primary basis for the analysis, and industrial engineers still respect time as "the only incorruptible basis for measurement." All cost elements are usually related to time in the industrial engineering analysis process. The new simulation tools and the earlier operations research techniques all use a time scale as the basis for comparative analysis of functions and costs.

The main weakness in the industrial engineering analysis occurs when indirect costs are applied. In most cases, these data are supplied by the enterprise's accounting department, which usually continues to use "distribution" type cost analysis techniques to calculate burden and overhead rates and apply them to product cost. The result is a distortion in the computation of costs, because they reflect production volume variations and tax rules dominated burden calculations.

The industrial engineer also approaches the definition of the manufacturing system in a bottom-up manner. In most cases, the industrial engineer begins his work with the analysis of the product and its components to define the best process and/or material for production. The engineer often applies the techniques of value analysis and group technology to achieve the minimum product cost through planning and design. This often results in feedback to the designer and modifications in the design which simplify the process and reduce

An Industrial Engineering Approach

the product cost. Changes can include the choice of materials, the shape of the part, the required tolerances, the method of manufacture, or the substitution of prior or standard parts. These early cost reduction efforts are usually production labor time related, but also involve consideration of material cost and overhead expense based on the choice of process equipment.

As a part of the product analysis process, the industrial engineer develops the "manufacturing bill of materials" (MBoM). The MBoM is structured to show the sequence of the assembly and processing of the product. This document differs from the "engineering bill of materials" and the "parts list" or material list in that it shows the step-by-step sequence of the development of the product.

The engineering bill of materials is usually design discipline oriented, and the parts list or material list is usually vendor or component function oriented. Only the manufacturing bill of material is related to the process. It is used as the basis for production scheduling and process planning.

Figure 3.1 shows the typical structure of a manufacturing bill of materials and the level-by-level control or breakpoints in the manufacturing process.

SIMSCODER® INDENT LEVEL CODE	INDENT LEVEL DESCRIPTION
1XX	1) FINAL ASSEMBLY - PRODUCT (*OR CONTRACT*)
2XX	2) MAJOR SUBASSEMBLY (*OR PRODUCT*)
3XX	3) SUB-SUBASSEMBLY (*OR SUBASSEMBLY*)
4XX	4) DETAIL PART (FINISHED)
5XX	5) DETAIL PART (SEMI-FINISHED)
6XX	6) RAW MATERIAL
3XX	3) SUB-SUBASSEMBLY
4XX	4) DETAIL PART (FINSIHED)
6XX	6) RAW MATERIAL
7XX	7) FASTENERS
2XX	2) MAJOR SUBASSEMBLY
3XX	3) SUB-SUBASSEMBLY
7XX	7) FASTENERS
3XX	3) SUB-SUBASSEMBLY
7XX	7) FASTENERS

Figure 3.1 Typical manufacturing bill of materials structure.

These breakpoints in the process are control or inventory points where the work-in-process changes its characteristics and identity. These points also define the "indent" levels of the manufacturing bill of materials. Figure 3.1 shows a typical "indented" manufacturing bill of materials format as it would be written in a manual system. A computer can also present it in the same type of format if desired.

This format demonstrates another element of the bottom-up approach to management; the structure of the product is the dominant element in the definition of the control points or data access points in the management process. Costs can be collected at these process change points and the work-in-process inventory can be valued at each level of completion. This permits a cumulative cost computation and detailed part and operations cost definition.

This bottom-up philosophy permits the industrial engineer to approach cost analysis and cost accounting in an incremental or level-by-level manner based on a precise definition of each cost element and a build up to a total product cost.

B. THE SCIENTIFIC APPROACH TO PERFORMANCE MEASUREMENT

The scientific method is also a bottom-up approach. The term "scientific management," popular in the 1930 to 1950 era, is probably obsolete and inappropriate in the modern industrial environment. However, the "scientific approach" remains a sound basis for analyzing and solving many problems, whether technical or managerial. It requires gathering data, analyzing and interpreting the data, developing conclusions from the analysis; and developing formal conclusions and recommendations based on an objective review of alternative solutions. This is the process that industrial engineering has historically used.

To provide a basis for performance measurement, it is obviously essential to define the objectives. This is a common bottom-up industrial engineering task. Product designs and product performance specifications define the product; product performance requirements define the quality standards; and the market forecasts and production schedules define the required quantities. Profit objectives define the economic targets. A scientifically developed standard cost system and product budget defines the manufacturing cost targets and provides the basis for evaluating cost effectiveness.

Today's performance measurement methods evolved from a bottom-up "scientific" analysis process supported by a great deal of shop floor experimentation as industrial engineering established itself as a profession. In the development of a scientific approach to manufacturing cost accounting and estimating, it is therefore necessary to start with the measurement and definition of the elements of cost. These are labor, materials, overhead or burden, and general

An Industrial Engineering Approach

and administrative expense. If each of these can be related to the processing of each product unit on a time use basis, it is feasible to develop precise product and process operation cost definitions. *The cost definition should be based on the time use of facilities and labor plus the net use of the materials consumed by the product and in supporting production.*

The resulting cost definition should be developed for each detailed part on a step-by-step operation basis. It should provide the required cost elements for the computation of a precise "landed cost of sales" for the product. It should allow detailed analysis of individual operations and precise definition of every cost element and variance. This level of precision and detail is needed to make informed manufacturing, design, and marketing decisions in today's capital intensive manufacturing environment and competitive global market.

In the development of a performance measurement system, the critical criteria are the achievement of product quality and production quantity objectives with cost effective performance leading to profitable operations.

C. WORK MEASUREMENT METHODS AND STANDARD TIMES

To develop a standard cost for a manufacturing or materials handling operation it is essential to first define the operation and the standard work cycle. This requires a detailed work study to define the elements of the task, and a means to measure the work cycle and time consumed by the operation and by each of its work elements. In planning a new process or operation, it is also essential to accurately estimate the time which will be used by the new task or process. This usually requires the availability of predetermined work time standards.

The concept of using work measurement and work simplification as management tools can probably be traced to Eli Whitney's development of the cotton gin and interchangeable musket parts in the eighteenth century. He surely recognized the labor saving impact of repetitive operations and the avoidance of "cut and fit" assembly. But the techniques of work measurement began to be recognized as a professional discipline with the work of Fredrick W. Taylor and Frank Gilbreth in the early part of the twentieth century. Taylor pioneered the concept of work simplification and Gilbreth defined work element analysis with his "Therblig" system of motion study. Gilbreth really initiated the concept of predetermined time standards which were later developed into "methods time measurement" (MTM) by Maynard, Stegmerton, and Schwab.

This predetermined time approach allows the analysis of tasks and the estimation of the time required to perform a new operation before it actually occurs. The detailed time elements are developed by the averaging and/or statistical analysis of a large body of historical data, often based on many stopwatch measured work cycles. Many practitioners and firms have further developed the concepts of predetermined time standards for specific, and in some cases more general, applications. More recently, the Maynard organization has been promoting "MOST" (Maynard Operations Sequence Technique—Kjell B. Zandin, 1980) as a simplified predetermined time standards application tool. Time standards have also been compiled into computer data bases for use in the computer generation of work standards, the development of work schedules, and the simulation of operations. Qualified practitioners can achieve an accuracy of $\pm 5\%$ in the development of task standards based upon the use of such predetermined time standards as MTM and MOST.

Today there are several basic techniques which are commonly used to measure work time in industrial operations. They are all based on the relationship between an analysis of the task and the measurement of time.

The time to perform work varies with the assigned worker and the method of operation. It is, therefore, necessary to establish a basic or standard work method or task definition and to normalize the expected or measured time required to do the work in order to neutralize the human variations and establish a usable work standard.

The most basic work measurement procedure uses some form of stopwatch time study. This method requires the personal observation of the operation by the time study engineer or technician, or the use of a film or video camera with a timing device for later office analysis. The most flexible and least expensive procedure is the work sampling method. Work sampling is based on a statistical analysis of random multiple observations of work activities. In either case, the results of the study can provide the data base for a predetermined time standard system. In all cases, the volume of data, the scope of the studies, and the skill of the analyses will determine the degree of precision or predictablility derived from the standards.

In recent times, the use of predetermined time standards has gained favor for work planning and shop scheduling. Predetermined time standards are based on data derived from many detailed time studies and historical data. The standards are usually developed by synthesis of the operation or from stopwatch or work sampling data. Time and task standards are fixed by the work measurement study and can be retained and reused repeatedly until methods or products change. Random errors should be expected, but skewed error patterns will imply erroneous data. This will indicate the need for further study. Predetermined time standards have also been applied in the development of wage incentive programs.

In applying any of these techniques, and before beginning the actual timing of an operation, the observer should study the operation's work cycle to define and become familiar with the operation's labor elements and record them on the time study document. The labor elements should be small enough to permit detailed analysis of the operation, but also large enough and definable enough to be accurately observed and timed with a stopwatch or camera.

The work measurement practitioner must simultaneously "rate" the operator's performance during the timing of the operation in order to develop a valid standard. "Rating" is the evaluation of the operator's speed or work pace while being timed as compared to a "normal" or expected average performance for workers performing that type of task. This is a subjective evaluation and it can cause some of the later recognized variances in time standards.

After an average or modal time is developed from the accumulated observations, and rated for each work element, a "standard" time can be established.

Despite all the system based approaches, the most established method for work time measurement continues to be the stopwatch study. However, the accuracy of the stopwatch study largely depends on the skill of the time study practitioner, the quality of the analysis, and the rating of the work cycle or pattern. The stopwatch time study technique involves actual observation and on-the-job stop watch timing. It has an intrusive impact on the worker and can cause labor relations problems. It is not the most precise technique, mainly because of the subjectivity of the rating procedure. Stopwatch studies can generally produce standards within a $\pm 10\%$ accuracy. It is also probably the most expensive, and when studying a non-cyclic work pattern, probably the most difficult technique to use.

There have been many technical developments in the design of the time keeping and recording equipment to make stopwatch studies more precise and the timing devices easier to use. Some practitioners use film and video cameras with built-in recording clocks. The accuracy and cost effectiveness of time study has been enhanced by the use of these cameras and more precise timing devices. This permits the "desk study" of the operation and simplifies the shop floor observation process. It is also less intrusive and less likely to cause labor relations problems.

Some have tried to measure work elements to the one hundredth, and even to the one thousandth of a minute. This attempt at minute measurements appears to be unrealistic. One cannot achieve "precision by division." In the final analysis, the validity and accuracy of a stopwatch study depends on the skill of the practitioner, the size of the sample used in the study, the cooperation of the operator(s) being studied, and the quality of the statistical analysis of the data to develop a valid and reliable standard. The reliability and consistency of the time study engineer's rating is a primary factor in determining the validity of the standard.

Work sampling is a valid, inexpensive, and statistically based means for studying the labor content and random elements of an operation. It is particularly useful in dealing with non-cyclic operations. Relatively inexperienced people can perform the observation function in a properly structured work sampling study since the observation does not usually require subjective rating of operator performance.

The concept of work sampling is based on the laws of probability. This means that a small number of chance occurrences tends to follow the same activity distribution pattern that a large number produces. In applying this statistical technique to work sampling, the number of times an operator is observed performing a particular function or work element will relate to the total observations in the study in approximately the same percentage or ratio as the total time required by the operation or work element related to the total labor time worked. The more observations, the more accurate will be the results. Also, the more detailed the study, the more observations will be required to assure and maintain statistical validity. The same work pattern analysis and detailed operation breakdown must precede the sampling study as in a stopwatch study. Performance rating is also often applicable in the establishment of work sampling based standards.

An advantage of the work sampling technique is that it can be used to study staffing, labor performance, events, methods, clerical activity, materials handling, product related functions, or random activities. The study can be designed to collect data in all of these areas at the same time and each can later be separately analyzed.

A random schedule must be established for observing each element of the study. The schedule should vary from day to day and the sequence for observing the operators and the time intervals at which they are observed should be completely random. However, one feature of non-randomness is essential: every person, operation, event, or situation must be observed (or its absence noted) during every cycle of observations. This will result in the same number of observations for each operator or event, and will thus provide a valid basis (total observations) upon which to compute the activity percentages and develop the time elements. The total number of times that each activity element is observed is divided by the total observations to derive the percentage of the total time used by each work element. This percentage is then applied to the total available time and divided by the number of production units to develop the standard time per unit of production for each activity.

There are those who question or reject the use of work sampling as the basis for establishing work standards. Their reasoning is based on the likelihood of achieving a less accurate standard than is possible by using modern time study methods. However, the use of work sampling is quite effective in long cycle, and highly complex operations such as ship building, heavy machinery construction, non-cyclic assembly, and other non-serialized manufacturing operations. In such cases, work sampling can be much less costly to apply and just as adequate for the development of standards for cost projections and estimates, and the scheduling of production. Work sampling is not usually accurate enough for use in the development of wage incentive programs.

Historical methods for developing time standards are occasionally used. The standard is determined by dividing a particular period of time by the number of occurrences of the element in that period. This is not a very accurate technique because the time period usually includes extraneous activities in addition to the work elements which are being studied. The resulting times defined by this method are sometimes termed "Average Time Standards." They are usually unsuitable for use as the basis for a standard cost system or a wage incentive plan.

In summary, it is necessary to develop a time based work standard for the measurement of manufacturing performance before a precision cost system can be established. In doing so, it is necessary to use some type of work measurement procedure.

Dollar valuation of a task should be the last step in the cost system development. The dollar value of a time unit (minute or hour) will vary with the wage rate and fringe benefit program, with the assignment of individuals to the task, and with the work station's "use rate" value. Therefore, the dollar value of a task is not a valid basis for a cost system until the task and the economic circumstances have been fully defined. Dollar values should be adjusted to suit the current situation at the time of application. Time and task standards are fixed by the study and can be retained and reused repeatedly until methods or products change.

D. MATERIAL COSTS AND STANDARDS

In addition to the costs of manufacturing labor, defined above, it is essential to precisely define and include material costs and overhead in computing the product cost.

There are three basic classes of materials in a manufacturing operation—direct materials which become a part of the product; indirect or contingent materials which are used in production, usually in a non-linear relationship, but do not become a part of the product; and overhead or burden materials which are used to support production operations but do not have a linear relationship to the manufacturing system. From an industrial engineering point of view, the measurement and accounting for each of these materials classes requires a different approach. To start with, let us define these materials classes as:

1. Direct Materials

Direct materials are the raw materials and purchased parts or components which become a part of the finished product and are shipped to the customer with it.

These would include raw materials such as metals, plastics, paper, fiber, glass sand, chemicals, welding rods, etc.; and purchased parts or components such as labels, bearings, fasteners, paint, electrical fittings, motors, engines, tires, light bulbs, etc. In many products, particularly food, cosmetics, health and beauty aids, toys, and consumer hardware, the package must also be considered a direct material and part of the finished product at the end of the production line.

2. Indirect or Contingent Materials

Indirect or contingent materials are the raw materials or manufactured products which are consumed by the manufacturing process, usually in a non-linear relationship with production, and do not become a part of the product. These would include foundry sand, chemical catalysts, abrasives, coolants, consumable tools, and work-in-process protection materials such as coatings, oil, paper, dunnage, etc. In some cases, pallets and tote boxes are treated as consumable tools; in other cases they are capitalized along with tooling. In some operations where heavy use of power, water, or fuel is required for production, these utilities are separated from the heating and lighting utilities and treated as an indirect material. This would be true of fuel in glass and ceramics operations, steel making, foundries, and heat treating, and electric power in plating and some heat treating processes. In situations where the products are picked from miscellaneous stocks and packed along with a multiple of other products for shipment, the shipping and packing material should also be treated as an indirect material of manufacture.

3. Burden or Overhead Materials

Burden or overhead materials support the manufacturing operation but do not directly or linearly relate to the manufacture of the product. These include stationery, office forms, computer supplies, sanitary and janitor supplies, maintenance materials, lighting and environmental control power, heating fuel, potable water, fire water, etc.

From an industrial engineering point of view, it is essential to relate all of the direct, indirect or contingent, and overhead or burden material items and their costs to either the individual product or a specific operation. In the case of the indirect and contingent materials which are not specifically chargeable to a particular operation or part, their costs can be distributed to the product through the time use of facility charge. Their cost can then be absorbed into the product cost in an equitable manner based upon the operation time at each level of production. The overhead or burden materials will also be distributed through the time use of facility charge. The direct materials are identifiable in their relationship to the product or part, and therefore present few problems in achieving cost definition objectives.

An Industrial Engineering Approach

The time use of facility cost distribution and allocation concept is based on the belief that a combination of the fixed or measured proportion of the facility utilized, and the time of its use by an operation or work center, provides an incorruptible vehicle for the collection of period and indirect costs into the operation or activity utilizing that portion of the facility. This technique also isolates the portion of the facility which is being used, and its share of the charges, from the unused portion of the facility which is an expense to the enterprise, but not chargeable to the operation, activity, or product. It also avoids the error of using a volume dependent and variable cost base.

Industrial engineers and accountants often disagree about the cost accounting treatment of these indirect and burden materials both conceptually and philosophically. While the industrial engineer usually seeks to take a standard unit of consumption approach, the accountant usually applies the distribution accounting philosophy. Since this book focuses upon the development of a total absorption standard cost system, we will use the standard unit of consumption approach in the materials discussion below. In our discussion of these issues, we will take these material classes in order.

The establishment of direct materials' design quantity standards usually requires some quantitative adjustments to assure the proper accountability for scrap, chip loss, "nest and gain," and shrinkage. Scrap and shrinkage losses are also a consideration in the consumption and control of indirect and burden materials. However, with a few exceptions, as in cases when the material is perishable, a controlled substance, or a very high value item, the scrap and shrinkage losses and salvage revenues for burden materials are usually buried in the burden materials budget. In the case of indirect manufacturing materials, these cost factors can and should be defined and applied to the product cost.

All other materials require a sound definition and a defensible basis for inventory valuation. The cost base must deal with all of the usage variables and address both the landed purchase price of the materials and the inflation or deflation of vendor and market prices.

Industrial engineers generally address these issues by establishing a period standard cost (usually one year) for each unit of each item of material to be used. In establishing the material unit standard cost, the engineer uses actual, historical, calculated, or quoted prices or manufacturing costs, and tries to interpret and anticipate predictable cost and price trends and changes. The standard is established by computing a mid-period standard cost (Figure 2.5) which generates random and balanced actual versus standard material unit cost variances. If the variances tend to be skewed higher or lower than a standard pattern over the span of the period, this indicates a need to review the situation and adjust the material and product cost standards to produce a more balanced and random variance pattern. The adjustment would be applied to the material quantity and cost standard, and the inventory ownership cost. The adjustment would also impact the total standard product cost.

The mid-period standard cost of the adjusted standard direct material quantity per unit of production is used to develop an estimate of the expected manufactured cost of the product. It is also used to develop the product cost budget. The materials' cost and consumption variances are allocated to a special material cost variance account. From there, depending upon the nature of the operation, these variances are applied directly to the cost of the product, or are used to periodically adjust the material cost standard, or are included in the burden or contingent materials cost allocation process.

Burden or contingent materials present a more difficult cost definition and cost allocation problem. The material usage is not always related to a particular process, work station, product, or department. The usage is also time variable and difficult to record.

For example, coolants may be fed to machines from a central system and used in varying amounts on different operations, parts, or machines. The consumed coolant materials are replaced at the central system. No workplace accounting is possible. Foundry sand, lubricants, abrasives, and such reusable consumable tools as drills, tool bits, and milling cutters also follow this use pattern. Often such contingent materials can be charged as designated burden to a cost center or department. This is true of foundry sand specifically chargeable to the foundry, but not to a specific product or work station in the foundry.

Some contingent materials, such as grinding wheels used on a thread grinder, may also be specialized to a particular work station, but not to a particular operation or product. The machine might be the only consumer of the special grinding wheels, but it might grind threads on many different parts for several different products. This complicates the desire to charge the cost of the wheels to the product.

Thus, although it is a simple matter to apply the adjusted standard direct material quantity and cost to the first operation in the manufacture of a part or product, it is not as easy to capture the cost of the indirect or contingent materials on an operation-by operation basis or for a particular part. In the system proposed by this book, the indirect or contingent materials will be applied to the operation via the time use of facility or "use rate" technique. This will be discussed in detail in a later chapter. Burden materials will also be charged to the product through the same vehicle.

E. EQUIPMENT AND FACILITY COSTS

One of the most difficult and often distorted cost elements in the development of manufacturing costs is the ownership cost of facilities and equipment. These costs include depreciation or rent, utilities, insurance, maintenance, heating, ventilation and air conditioning (HVAC), and a number of related expenses.

The depreciation life defined by the Internal Revenue Code may be very different from the realistic life of an asset. The financial manager may also elect to use a rapid write-off technique. As a result, if tax rules are used, a non-uniform and often unrealistic ownership charge could be developed for the asset.

For example, rapidly developing technology in computer based manufacturing may make the acceptable life to obsolescence of a machine or computer much less than allowable in the tax code. At the same time, a specialized building such as a rack supported warehouse may fit the tax rules as a "machine" with a shorter depreciation life than a conventional building.

This raises a question about the need to define the "true cost" of manufacturing equipment and facilities rather than the "reconcilable" indirect costs which fit into the financial and tax reporting structure of the enterprise. The basic philosophy of this book states that there is little reason for a fit or match between these different cost data presentations. This philosophy also states that truth and reality in the manufacturing cost analysis system are essential to sound manufacturing operations and product design decisions. From the manufacturing and operations management point of view, the accounting life of a real estate or capital equipment asset should be as close to the expected functional life of the asset as possible.

Modern data base computer programs allow the manipulation of a common data base into a variety of reports which can be focused on the needs of different audiences and provide a variety of decision criteria. The same indirect expense and overhead source data and data base can usually be used in both the manufacturing and financial analysis functions. Measured standard data and true variance reporting also make more precise manufacturing information available to managers.

Various methods of data application result in different data presentations and a variety of solutions or reports. Equipment and facilities' utilization cost variances, when treated as adjustments to financial reports, become a basis for evaluating manufacturing performance within the enterprise structure. These variances can also help in adjusting standard costs to more accurately represent actual performance.

Equipment and facility capacity utilization variances are not usually caused by manufacturing decisions. They are usually a function of sales effectiveness and the resulting scheduling of production. Conversely, operations variances are usually manufacturing based unless they are skewed or non-random because of a faulty standards definition.

In the proposed system, capacity based cost variances will be charged to the profit and loss statement as an adjustment to the enterprise's overall performance report. They will not be applied to the product cost base.

F. OVERHEAD ALLOCATION METHODS

While the "use rate" has been discussed in the text without explaining its characteristics, structure, or function, this omission has been intentional. The "use rate" concept is a simple idea. The details of its content need more complete explanation than the concept itself.

The "use rate" concept is based on the use of an internal "rental rate" for facilities and equipment. The system charges all related overhead, maintenance, and other indirect expenses into the account of the asset and rents the asset to the user or the manufacturing operation on a time use basis. It works like an automobile rental operation.

In renting a car, the rental agency establishes a daily or hourly rate (time use of facility standard cost or "use rate") which includes *all* of the expenses of the business (plus a profit). The "use rate" is defined on the basis of a predicted number of rental days. The rate for a compact may be $45 per 24 hour day or partial day, based on 200 rental days per year and a one year or 15,000 mile ownership life for the car, followed by resale. The rental rate, based on the forecast 200 day use, includes the cost of owning the car plus the cost of garaging, maintenance, washing, insurance, licenses, office staff support, supervision, general and administrative expenses, and profit.

The renter pays for the fuel (materials) and provides the driver (labor) for the trip (operation). The renter only pays the rental company for the time use of the car. If it is underused (less than 200 days), the agency absorbs the cost as a negative profit variance. If it is used more than 200 days, the agency receives a positive variance and an additional profit.

Each renter only pays for the time use of the car, no more, no less. The cost of the operation to the renter does not include the cost of any idle time within the 200 day schedule. The renter does not save from the agency's good luck in renting the car more than the planned 200 days.

As in the case of a car rental rate, the "use rate" is the time based daily or hourly charge for the use of a building, vehicle, or piece of manufacturing equipment. It captures all of the management, overhead, burden, maintenance, and other indirect expenses into a time cost rate for the use of the equipment or facility. The rate is based on the scheduled or expected use of the equipment or facility during the accounting period (i.e., one year). While a car rental could use a daily or hourly charge, manufacturing would use an hourly or minute-based time charge.

Like the car renter, the user or manufacturing operation is only charged for the time that the equipment is being used. The cost of idle time within the expected standard annual utilization time is assigned to the enterprise and charged to the profit and loss statement of the business. It is not charged to the product. Overuse returns a profit to the owner and underuse results in a loss of profit.

An Industrial Engineering Approach

The direct material cost used in the auto rental operation is paid by the renter when purchasing fuel and oil. Manufacturing also defines the direct material cost and assigns it to the product in a data collection process which is separate from the "use rate." In car rental, the renter is the direct labor item. In manufacturing, direct labor cost is also a separate data collection item.

Thus the "use rate" is a time based rental charge which collects and allocates all relevant indirect manufacturing costs to the product on an equitable time use of facilities basis. It restricts allocation of indirect expenses to the product to the time use of facilities share which the product actually earns. This eliminates the erroneous impact of production volume variations.

This time use of facility theory is basic to the manufacturing costing philosophy presented in this book. When applied, it leads to the ground rules listed below. Some financial people and conventional accountants may feel that these ground rules are arbitrary and even heretical. The rules are arbitrary! However, they are based on fundamental theoretical principles. These principles can be stated as follows:

- The distribution of indirect expenses, overhead costs, and general and administrative expense (G & A) to the product should be based on the time use of the management staff organization, physical assets, and/or facilities which are used to make the product and that cause and/or absorb these expenses.
- The product and the manufacturing function should not be charged with the cost of unused facilities and staff which could have been put into use by more effective marketing and sales effort. The ownership and indirect cost "use rate" charges for idle facilities and staff should be charged into the enterprise's profit and loss statement.
- The overabsorbed burden and indirect expense variances from standard "use rate" schedules which are caused by schedule and sales based overuse of facilities (as in overtime and weekend operations) should be credited to the profit and loss statement.
- The product should only bear its direct costs and its fair share of the indirect, overhead, and G & A costs which can be allocated to it through a "use rate" charge based on its time use of the facilities.
- The cost value of the equipment and facility "use rate" should be calculated in the same manner as a fully absorbed rental charge for the equipment or facility. Computation of the "use rate" should be based on the scheduled annual use of the asset with full consideration of maintenance down time and expected work schedules.
- The realistic expected life and straight line depreciation of the asset should be used. The tax based depreciation life or the legal tax life schedules should only be used in the computation of the "use rate" when they coincide with, or approximate, the predicted true life of the asset.

The detailed procedure for building the "use rate" charge and assigning costs to it will be discussed in a later chapter.

G. BUILDING THE PRODUCT BUDGET

All of this information funnels into the the Product Cost Budget. This budget is really an estimate of the expected manufacturing cost of the product as calculated from measured standard cost data or engineering estimates. Table 3.1 shows the format structure of a product cost budget which includes both the factory cost and the landed cost of sales. This budget includes direct, indirect, and overhead expenses for the product.

When setting the product cost budget, the engineer must first develop and document the process of manufacturing and assembling the product. In most cases, this will require the definition or design of the process, the development of process flow diagrams and operations sheets, the selection and/or design of tooling, and the design and measurement or synthesis of work cycles and operations times. The estimator must also define the net direct materials requirements and the type and volume of any indirect or contingent materials or tooling specific to the process.

In the absence of standard cost data, the designer or manufacturing engineer uses other methods to estimate the cost of the product. Some of these estimates are based on the personal experience of the estimator. Some are built up by retrieving historical data on the cost of similar prior products or parts. In any case, an estimate is only as accurate as the data upon which it is based, and the compatibility between the forecasted and the actual manufacturing process.

An experienced "guesstimate" can be as accurate as a calculated estimate if the estimator really knows the product and process.

However, such "guesstimates" are an uncertain basis for predicting the cost of manufacturing. Computed estimates which are based on tested and measured standards will normally provide a far better basis for management decisions.

The estimate or product cost budget is used as the target for process performance evaluation and control, and as the basis for marketing and pricing decisions. The time use of facilities, total absorption standard cost approach, which is presented in this book, provides a tool for realistic and accurate engineering estimates, process performance projections, and product cost budgets.

An Industrial Engineering Approach 47

Table 3.1 Precision Cost Accounting, Product Budget, Format Before Development of the Use Rate Costing Technique

LABOR COST (Time x Rate)
    DIRECT LABOR - - - - Productive activities
    INDIRECT LABOR - - - Supporting activities - Material handlers, etc.
    FOR NOTHING COST - - Overtime and night shift premium
        (50%, 10%, etc.) - Union Steward time
    FRINGES - - - - - - Social Security, Hospitalization - Unemployment Tax,
        Worker's Compensation, Pensions & Vacations

MATERIAL COST
    STANDARD MATERIAL (Ms) - - Design Standard
    SCRAP ALLOWANCE (Ms) - - - - Protection against bad work
    NEST AND GAIN ALLOWANCE (F ng) - -
        Effect of schedule on material useage
    OBSOLESCENSE AND SHRINKAGE ALLOWANCE - -
        Effect of spoilage and changes in requirements
    CONTINGENT MATERIALS - - Consumable tools -
        Foundry Sand   Emery and Abrasives
        Catalysts   Coolants
        Press Lubricants   Etc.
        (Could be charged as Burden Materials)

BURDEN COST (Cost per Period - Time use of Facility)
    HOUSING COST - - - Amortization on plant - Cost of insurance & taxes
        Cost of heating, light, and ventilation - Cost of maintenance, etc.
    EQUIPMENT COST - - Amortization of purchase cost -
        Cost of maintenance - Cost of power, supplies, etc.
    SUPERVISION - - - - Direct Supervisors and Foremen -
        Superintendents, Operations Executives, etc.
    STAFF SUPPORT - - - Setup men - Tool maintenance, sharpening, etc.
        Shop clerks - Security and fire guards - Personnel administration
    BURDEN MATERIALS - Stationery and toilet supplies -
        Maintenance supplies - Handling equipment fuel, etc.

FACTORY COST OF SALES

GENERAL & ADMINISTRATIVE COST - - Office rentals - Interest -
    Executive salaries and Secretarial cost - Public Relations -
    General Administration, Financial Control, etc.
DISTRIBUTION COST - - Warehousing - Transportation - Shipping -
    Packing, etc.

LANDED COST OF SALES

SALES EXPENSE - - Advertising - Sales commission - Travel expense, etc.

GROSS COST OF SALES

PROFIT (Before Taxes)

PRICE

4
Time Use of Facilities Product Costing

A. FREEDOM FROM PRODUCTION VOLUME VARIATIONS

The "time use of facilities" and "use rate" concept of cost accounting is based on the belief that the product or process should only be charged with the time use of those portions of the facilities directly or indirectly involved in the process of making the product. This approach allows the assignment of all applicable charges to the product through the specific cost generating sources and facilities (or activities) used in the manufacturing and distribution process.

This approach is similar to responsibility or "activity based costing" (ABC) since it assigns costs to products based on the cost's origin. It differs because the period, indirect, and overhead elements of the manufacturing cost are related to the facility and are defined and charged to the product based on the time use of the facility or work station. These cost assignments are not related to volume or units of production, or to a percentage of such cost elements as labor or materials.

The key to the time use of facilities, total absorption standard costing approach is recognizing that the product or part being produced should not be charged for the cost of unused capacity, or credited with the overabsorption of overhead.

The defined cost of the product is based on its proper share of the cost of manufacturing and distribution—no more or less. The cost definition does not depend on production volume.

As stated earlier, most generally accepted accounting procedures use contemporary production volume as the basis for assignment of the overhead and burden costs of manufacturing and distribution operations to the product cost. This results in an apparent product cost which varies inversely with changes in production volume. As volume increases, the apparent unit cost declines, and vice versa. This technique does not provide a stable base or datum against which to measure manufacturing cost or performance.

In precision costing, all charges are assigned to the operation, part, or product on the basis of the time use share or proportion of the available facility capacity used in manufacturing and distributing the part or product. The product is only charged for the time during which the physical facilities are used to make or handle the product.

All burden and overhead expenses and indirect labor used in the process are related to the product's time use of the facilities through the work station's "use rate" and the operation's cycle time. All hands-on labor is also related to the part or product based on the work cycle time. Material costs are related to the product or part on the basis of the design standard materials used plus the appropriate scrap and "nest and gain" adjustments.

This approach is generally immune to the impact of fluctuating production volume. The product cost definition is independent of the proportion of the plant capacity in use at any time. The cost computation includes only that portion of the scheduled capacity which is actually used in the making or handling of the specific part or product, and no more.

Obviously, when a facility schedule is changed from single to multiple shifts, or from five to six or seven day operations, the distribution of fixed and period facility and overhead costs must be altered. The computation of an area or space "use rate" for allocating fixed and period costs to the defined and fixed net productive space base will generate a new and different space cost or "use rate" standard for each of the plant's standard workday or shift schedules.

Accommodating the new schedule will require computation of changes in each work station's "use rate." This treatment of long term or permanent schedule changes is a reluctant but realistic concession to distribution type accounting practice.

This adjustment to permanent or period schedule changes is separate from the computation of variances based on normal cost fluctuations. Short term variations in the shift and work day schedule should be treated as variances.

Beyond the impact of shift changes on work schedules, the normal drift of costs in the economy will also require periodic adjustments in the "use rate" standards. The standards should be reviewed and updated at least annually, and in volatile economic periods, possibly every six months. The magnitude and trend of these adjustments must be recognized in the product costing procedure and in marketing and pricing decisions.

Together, all of this "use" information establishes the base for achieving the first three objectives of a precision manufacturing cost accounting system.

The three primary objectives of the cost system are:

1. To provide management with a reliable tool for measuring and managing the performance of the manufacturing and distribution operations.
2. To determine the actual cost of a product as compared to pre-manufacture cost estimates and market prices.
3. To provide information for designers who design to cost and apply value engineering principles in product development.

Retrieval of prior costs from similar parts and products is an essential part of the product cost estimating process and can help in making parts standardization decisions.

To achieve these cost management goals, it is essential to establish a stable baseline or standard against which to compare actual performance. This base must be independent of current, future, or past variations in production volume. The most practical, stable, and incorruptible cost measurement method must also be defined.

The cost standard must specify a time period (such as a year or half year) and be revised for each period to accomodate inflation and facility scheduling changes. To provide a benchmark for performance measurement and evaluation, the standard must be stable for the duration of the defined standard period. Performance deviations are measured as variances from the period standard.

B. DEVELOPING A STABLE COST BASE

The requirement for a stable measurement standard is the philosophical basis for all standard cost systems used in cost accounting. However, the problem with most conventional, and some special, cost accounting systems is that they build their standards on a variable base.

Most of the standard cost systems which apply generally accepted accounting principles distribute burden and overhead expenses to the product based on the volume of production or as a percentage of labor or material cost. In each of these methods, the basis for the cost distribution is a variable and the product is charged with unused or unrelated facility burden and overhead. The resulting calculation of the operation's costs are inaccurate and variable.

A stable and reliable basis for detailed manufacturing cost measurement and control is needed. It must be free from the effects of production volume variations and also include all of the operating costs of the manufacturing and distribution system.

All burden and overhead costs of a manufacturing business can be related to the two genuinely stable and independent elements of the cost structure.

These are the time consumed by the operation cycle and the net operating space occupied by the work station and equipment. The time use or consumption of these stable measurement elements varies linearly with the operation's cycle time. Time is the only truly reliable standard of measurement for manufacturing and distribution operations.

A square foot of net operating space is a stable vehicle for collection and allocation of indirect and overhead costs to work stations and their production operations. In this book, the definition of "net operating space" will be: The space actually utilized by the manufacturing and/or distribution operations and their supporting aisles.

The net operating space does not include toilets, locker rooms, stairwells, in-plant staff and executive offices, meeting rooms, cafeterias, medical facilities, utility areas (power and HVAC plant, etc.), maintenance shops, guard posts, etc. In warehousing operations, shipping and receiving dock areas may also be treated separately.

The ownership and operating costs of these excluded areas are included in the total annual building or facility ownership and operating cost. This total facility cost, plus such other costs as insurance, taxes, utilities, and maintenance which are associated with facility ownership, will be divided by the area of the net operating space to develop an annual facility ownership cost per square foot of net operating space. This annual cost per square foot will be divided by the scheduled operating hours of the enterprise to produce an hourly space "use rate" in terms of cost per square foot, per hour. This cost figure is multiplied by the area (square feet) occupied by the work station and its supporting aisles to develop an hourly *space* "use rate" for the work station. This is the basic vehicle for the accumulation and assignment of burden and overhead expenses to the work station and the product.

In a mathematical format this would be:

$$U_{WSS} = \frac{[B_A/S_{NO}]}{H_{SO}} \times S_{WS}$$

Where:
U_{WSS} = work station space "use rate" per hour.
B_A = the annual cost of owning and operating the buildings.
S_{NO} = net operating space of the plant facility.
H_{SO} = standard operating hours of the enterprise (i.e., 2080/yr).
S_{WS} = work station space in square feet including supporting aisle and maintenace clearance.

This "use rate" cost base replaces such variable burden and/or overhead bases as labor time and material cost. It will permit the development of a space cost "use rate" or module which can also be used as a vehicle for the capture and assignment of indirect or overhead expenses generated by such other cost sources as the production machinery and equipment, the general and administrative expense, overhead and burden labor, and burden materials. This procedure allows all overhead and indirect expenses to be directly assigned to each operation in an equitable relationship to their contribution to the operation and, at the same time, independently of the overall volume produced.

The "use rate" for the work station's production machinery and equipment is calculated from the equipment depreciation or lease rate ownership cost and the directly attributable operating cost of the equipment. The equipment "use rate" is then added to the work station's area "use rate." The method of computing the equipment "use rate" will be detailed later.

The cost or charge for utilization of the work station is linearly related to the cycle time of the operation. This relationship is similar to that of renting and using automobiles or construction or garden equipment. One only pays for the time that the equipment is in one's possession and use. The time use of the work station (and its related burden and overhead expenses) is the same as the cycle time of the operation being performed on the work piece or assembly. The cost or "use rate" for using the work station is the same for each hour or minute it is used while performing the operation. The product is only charged for the minutes used.

The work station is treated as an "activity," to capture the "activity cost" based on the work station's "use rate."

The unused or idle time of the work station should not be charged to the product or operation. This time is not involved in the work cycle. It is a cost to the enterprise but not to the product. The cost assigned to the part or product is computed from the time use of the facility consumed by each unit of production. This time use cost is independent of the total volume of production and any variations in the total volume.

At the detail level, the "use rate" approach meets the needs and criteria of cost accounting, manufacturing operations management, engineering estimating, and product design. At a higher level, the manufacturing bill of materials and the part coding system must also be designed to support the manufacturing management information system and the precision cost accounting system.

C. LEVEL-BY-LEVEL COST ACCUMULATION

The manufacturing bill of materials (MBoM) is structured to reflect the levels of manufacturing and the assembly sequence (Figure 2.1). Each indent level of the bill of materials defines the level of completion of a component or subassembly as it progresses through the manufacturing process.

The manufacturing bill of material is a key document in the development of the materials requirements plan (MRP) and the manufacturing process. It also provides the basic information structure for material (inventory) control and the overall management of the manufacturing operation. Its indent levels and part coding structure are the basis for level-by-level inventory management and cost control.

As each part or item passes from one process or indent level to the next, it changes its identity and collects cost. As the item passes from one level of completion to the next it is "posted" or recorded into the work-in-process inventory with its new identity and accumulated cost. In many processes, the part or product component never touches the floor as it moves to the next operation. Nevertheless, its accumulated cost and its change in identity must be recognized and recorded in the work-in-process (WIP) inventory as it is tracked through the manufacturing and distribution processes.

Each component or part enters the WIP inventory at the cost accumulated from all of the previously completed operations. Each part or sub-assembly passes through a work-in-process inventory level or "posting" point at each step in the process. As the identity and value of the part or component changes, the total value of the work-in-process inventory changes.

For example, a piece from a bar of steel could be drawn from the raw materials inventory at the landed cost of its purchase plus the cost of cutting it from the mill bar. It could then be machined into a spur gear blank. At that point its identity and part number would change from bar stock to spur gear blank. The combined time based cost of the machining labor and the work station "use rate" would be added to its material cost as the gear blank is posted into the work-in-process inventory.

The part would be inventoried as a spur gear blank at the cost accumulated to that point in the process. At the next step, the spur gear blank would be drawn from inventory at the accumulated work-in-process inventory cost. It would be treated as the material and entered into the next operation to be hobbed into a spur gear. The spur gear blank would disappear from the inventory and the finished spur gear would be inventoried with its new identity. It would add the labor and "use rate" charges from the hobbing operation to compute its new inventory value.

At each step or level in the above described example, the standard labor cost and the standard work station "use rate" will be used to compute the inventory book value of the part or component.

There are several uses for the costed manufacturing bill of material. If the manufacturing bill of material is standard costed at each level, it can be the basis for the level-by-level product costing procedure and the product cost estimate. The resulting cost estimate also provides the basis for measuring performance at each stage of the process.

The costed bill of materials will also provide the data for pre-production estimating of the total product cost and the cost of individual parts. The accumulated total product cost, and the finished parts and sub-assembly costs are extracted to produce the product cost estimate.

The standard cost should be used as the basis for costing each level of the bill of materials and for estimating the product cost. If the variances are continuously skewed in the same direction, the standards should be reviewed and either the process or standards, or both, should be corrected.

The level-by-level inventory value should also be computed in two ways. The standard time should be applied to the standard "use rate" to assign the standard facility cost to the part and operation. In addition, the actual time should be applied to the "use rate" of the actual machine or work station used, even (and especially) when it is not the work station specified on the operation sheet or work schedule. The actual labor cost should also be computed on the basis of the rate actually paid to the worker who performed the task, as well as the standard rate if it is different.

The work-in-process inventory record should be designed to show both the standard time based standard cost of the part or component and an actual time and standard cost based "actual cost" for use in the computation of performance variances. When computing the "actual cost," the actual employee labor cost and the actual machine or work station "use rate" should be used along with the actual time. If a different worker, machine, or work station is used as opposed to the standard shown on the operation sheet, this will show a variance which must be identified and costed. If a different pay rate is charged, this must also be noted.

For example, if the cycle time is standard and a worker with a high pay grade is assigned to a machine which commands a lower wage, it is customary for the worker to be paid at his regular rate. This would increase the product cost and show a plus labor variance for the operation. It would probably not show any variance in the total payroll account. Conversely, if a lower paid worker is assigned to a work station that commands a higher rate, the worker would usually be paid at that higher rate. In this case, if the time is standard, the operation cost would be standard, but the payroll account would show a plus variance for the worker's pay. If the operation is performed in the standard time on a machine that has a higher "use rate" than that specified in the operation sheet, the part cost would show a plus variance in the "use rate" and probably in the labor rate as well.

The computed level-by-level costs can be used to compare standard versus actual activity at each stage of the operation. In each case, the use of standard element costs and comparison with the standard methods and costs will allow a detailed management analysis of the operation and correction or acceptance of the variances.

In these ways, the costed manufacturing bill of materials provides the basis for estimating or defining the cost of the product. It is the information framework or format upon which the product budget or cost estimate is built. The costed or priced manufacturing bill of materials is a product cost estimate or budget.

D. COST RETRIEVAL FOR ESTIMATING AND DESIGN

Very few products are designed from scratch and then immediately put into production for the market. Most products evolve from prior versions or prototypes which are modified and improved for better performance and marketability. This process requires the retrieval of prior component designs for evaluation, modification, or replacement. This also requires the retrieval of prior manufacturing costs for comparison with the new design proposals.

Application of a continuous product improvement policy requires an incremental design and development effort. This requires the ability to retrieve and evaluate prior product design details and manufacturing costs. It is also necessary to define and recover the cost of each operation used in the manufacture of the prior design in order to evaluate the benefits and the cost effectiveness to be derived from the recommended design changes in the new product, part, or component.

The pre-manufacture product cost estimating process also requires detailed product design information and retrieval of manufacturing methods and cost data. It is essential to recapture prior design and manufacturing costs as an input to the cost estimating process. The detail cost data must be retrievable on a product, sub-assembly, part, and operation basis in order to make it useful in the estimating and process design effort.

As stated earlier, one of the primary purposes of a cost accounting system is to "determine the actual cost of a product as compared to pre-manufacture cost estimates and market prices." For this reason, the cost data must be retrievable on both a standard cost and actual cost basis. These data will also help the designer to utilize prior manufacturing experience in the redesign of the product or parts to achieve more cost effective production.

The concepts of a numerical and level-by-level manufacturing bill of materials and a matrix type part number coding system are designed to provide detailed manufacturing, product and cost information retrieval. The structure of a level-by-level numerical manufacturing bill of materials permits data extraction in either a vertical or horizontal mode. These techniques will be discussed in a later chapter.

Vertically, the product code accesses a complete bill of materials for each product. Upward vertical access using a part number and scanning of all of the bills of materials in all of the product lines will provide a complete "where used" listing for the part. This process is essential for development of parts standardization across product lines. It also allows the designer or estimator to retrieve design and cost data from prior uses of the part or component in other products of the company.

Horizontal access to the level-by-level numerical bill of materials permits scanning of all parts or components at a given level in the operation using the indent level code. For example, all semi-finished parts, all finished parts, all sub-assemblies, all fasteners, etc. can be retrieved. This scanning identifies similar or common use parts, semi-finished parts, purchased items, fasteners, etc. for possible reuse or retrieval of cost data.

These procedures allow the designer to identify similar and/or adaptable parts and components which can be used to enhance standardization in the new design. These procedures also allow the retrieval of prior use part costs for reference or inclusion in the product cost estimate.

In each of these procedures, the part or component identification code or part number is retrieved and listed in the new product bill of materials or the material and part listing for the cost estimate. Having identified the desired prior use part or component and its identification number, the user (or computer) accesses the work-in-process inventory and retrieves the posted standard cost and the latest or average actual cost for the item. This cost information is then used in the design evaluation procedure and in the development of the product cost estimate. This retrieval process can also help in the consolidation of common parts inventory to reduce work-in-process and/or spare parts service stocks.

E. PRECISE PART/OPERATION COSTING AND ESTIMATING

The depth of data this numerical bill of materials and matrix coding system contributes is vital to fulfilling the stated purpose of a precision costing system. It provides the access capability for accurate measurement of actual performance and the development of more precise part and operation costing and estimating.

The stated purpose of a precision costing system is to achieve a capability for accurate measurement of actual performance and the development of more precise part and operation costing and estimating. To accomplish this, it is necessary to include all of the elements of cost in the procedure and data base. This requires the definition of each of the cost elements and the design of a data base and collection procedure to capture the cost details.

The time use of facility approach to cost accounting requires the establishment of a module or unit of cost which is related to a stable element of the facility.

In the following chapters, we will discuss and identify the elements of cost which are to be collected. The "use rate" concept provides a medium for accumulating the results of the collection procedure.

5

The Level-by-Level, Total Absorption, Standard Cost Concept

A. THE TIME USE OF FACILITIES—"USE RATE" APPROACH

Earlier chapters have dealt with some of the specific problems inherent in the general and cost accounting practices used by most manufacturing and distribution companies. These include the methods commonly used to allocate indirect, burden, and overhead expenses, the methods for dealing with idle facility costs, the handling of general and administrative expenses (G & A), and the problem of pricing in relation to actual product cost.

Conventional accounting practices, which distribute all costs to current production, also charge idle facility capacity cost to the product. But idle capacity has no role in the manufacture of the product! The cost of the idle plant capacity should not be charged to the product: it should be recognized as an overhead expense and an adjustment to profit, not as a product cost.

At the same time, that portion of the enterprise overhead which is designated as general and administrative expenses is also usually left out of the product costing calculation and is assigned as a below the line deduction from gross profit. A proper share of G & A should be included in the product cost.

These procedures miscalculate the true cost of the product and penalize the manufacturing and distribution operations for sales weaknesses over which they have no control. Some manufacturers even overproduce to absorb overhead in the false assumption that this reduces their product cost.

The most serious problem is conventional cost accounting's development of an inverse cost relationship between production volume and inventory valuation, because these indirect expenses are all fully charged to current production. In periods of declining sales, the apparent cost of the product rises, bringing suggestions of price increases in the face of weak sales performance. In good sales periods, the apparent product cost declines, suggesting either a lowering of prices or higher profits. Neither inventory valuation reflects the true cost of manufacturing the product.

It is time to rethink our cost analysis practices and to develop more responsive and precise techniques for costing manufacturing operations and products. Pricing will always be market dependent and only indirectly related to cost. The use of time use of facilities-precision costing techniques and "use rates" for allocation of indirect and period costs will more accurately define actual product costs.

The primary difference between "use rate" cost accounting and conventional cost accounting methods is in the means for applying indirect and period costs to the product. The "use rate" cost accounting procedure applies the capital investment burden (depreciation), manufacturing burden and overhead, indirect costs, and general and administrative expenses to a stable, predictable, measurement unit (i.e., square feet and time) on a time use of facilities, total absorption, standard cost basis. The time use of labor, tools, machinery, and plant facilities provides measurable and definable standards for each operation. These standards are a constant for each unit of production and/or operation in a product's manufacturing process. They can be charged to the product, operation, or responsible operating department, in a clearly definable manner.

Conversely, conventional cost accounting usually distributes the total current overhead via a variable burden vehicle (material cost or labor cost) as a volume related (variable) percentage multiple of that burden vehicle's cost. The product or operation labor time (cost), or the product's material cost, is often used as the basis for distributing these overhead charges to the product or operation.

Production volume based cost allocation is basically unsound. Burden and overhead costs are usually fixed (or nearly stable) and do not normally fluctuate with production volume or labor or material costs. They should not be applied to product cost via a variable burden vehicle. Those indirect costs which do vary in parallel with production can usually be identified with, and be related to a particular operation or work station. In that case, they can be included in the computation of that work station's "use rate."

The process of charging costs to the responsible operating departments or specific products was discussed earlier. The primary production volume sensitive variable in the precision costing "use rate" structure is the definition of that portion of the plant's manufacturing capacity and operating time which is scheduled and actually used to support the operation and/or the production schedule. This definition can be made on a period standard basis. It should be periodically adjusted to accommodate long-term or permanent changes in the work schedule.

The time use of facilities or "use rate" technique corrects the variable product valuation problem and offers a better basis for allocating the actual accrued cost share of facilities, burden, overhead, and direct expense to the product. The time use of labor, machinery, and plant facilities provides measurable and definable cost standards for each operation or "activity." These standards can be used to generate realistic cost data which are not distorted by variations in sales or production volume. These costs can be accrued into the product's inventory valuation on the basis of operation time through the "use rate" procedure. Idle plant capacity, and facilities used for other production, have no role in the manufacture of the product and no effect on the cost of the product.

In the precision cost context it is essential for the "use rate" to include all related costs for absorption into the product cost. This will allow the inventory valuation of work-in-process or semi-finished parts to be correctly defined and recorded at each level or step in the manufacturing and distribution process. The absorption of these costs should include their proper share of the general and administrative expense and no more than the product's valid and appropriate time use share of facility costs.

The time use of facilities technique and "use rate" procedure provides a baseline from which to compute acceptable margins for profitable and competitive market pricing. These techniques also provide a better basis for responsibility accounting. Cost collection and analysis can be focused on specific work stations, departments, products, parts, operations, workers, and markets.

The techniques used to support the design-to-cost and engineering estimating operations were discussed in the presentation of the level-by-level approach to parts costing in the previous chapter. These procedures are also essential when costing parts which are to be standardized or used in something other than their original product. Precise cost figures are also needed when products are to be "engineered to cost" for future production in support to a price sensitive market. Whether tracking production parts, semi-finished parts to be altered for multiple application, costing items to be sold as spares or service support parts, or valuing work-in-process inventory, the engineered precision cost approach applies.

The principles of engineered precision costs can be simply stated as follows:

- Each performance element or "activity" in a manufacturing, warehousing, transportation, or merchandising operation involves the use of facilities and equipment, the use of materials and/or supplies, the expenditure of labor, and the use of money.

- By using industrial engineering work measurement techniques, the direct and variable cost elements of labor, material, and supplies can be defined, computed, and predicted for each operation and/or part with a reasonable degree of accuracy.
- Indirect expenses and burden and overhead charges can be defined and collected into a time and space based work station or facility "use rate."
- Indirect or overhead charges can be assigned to the cost of operations in direct and equitable proportion to their (the costs) time contribution to each function or "activity" by applying a time use of facility approach and a work station "use rate."

In summary, as outlined in Chapter 4, the basic concept of the time use of facilities or "use rate" approach is that all of the manufacturing overhead expenses are distributed to the operating elements of a business on the basis of each operation's actual time use of the capital facilities of the enterprise. Since the area of a building is a stable base through which the overhead can be allocated, indirect expenses and charges can be allocated to operations or "activities" on the basis of the actual time use of particular facilities (such as manufacturing space, storage space, office space, and manufacturing and materials handling equipment, etc.).

The time use of facility method for the application of indirect and overhead expenses to the product helps to solve the cost inaccuracy problems caused by the conventional accounting practices. The time use of facility "use rate":

- Establishes a rental rate or "use rate" for each work station.
- Includes all of the indirect, burden, and overhead costs which are attributable to the work station.
- Is charged to the using product and operation just for the time used during the actual operation.
- Is based on the standard scheduled annual hours of use of the work station.
- Defines the impact of schedule or performance based under-or overutilization of the facility which is not usually considered a valid product cost; this cost is treated as a variance and an adjustment to profit.
- Adds charges to the labor cost for the work cycle time of each operation, and the total operation cost is added to the material cost to compute the level-by-level inventory valuation at each level or step in the manufacturing process.

The acceleration of capital intensity and the decline of labor content in manufacturing and distribution operations makes the need for complete allocation of indirect expenses more acute.

B. THE TOTAL ABSORPTION STANDARD COST APPROACH

In practice, complete allocation of indirect expenses to the product is often referred to as "total absorption" costing. This type of costing is based on the "bucket concept" of cost data collection. It assumes that each part flowing from one operation to the next in the manufacturing or distribution system has a "bucket" attached to it. That "bucket" is really the costed work-in-process (WIP) inventory record of the item.

The objective of precision, total absorption, standard costing is to calculate correct and complete valuations of product and operating costs at every step in the manufacturing process. All elements of labor, material, burden, overhead, and general and administrative costs are accumulated into the product cost on a cumulative time use of facility basis as they are generated at each step in the process. The shares of costs which are allocated to the product are not related to the volume of production or the capacity utilization of the facility. The total standard and actual accumulated cost of the part or product at any point in its flow through the process is available by looking into the "bucket" or the inventory valuation of the work-in-process at that point. If the volume based distributive accounting philosophy is allowed to dominate the cost accounting procedure when assigning overhead and indirect expenses to product costs, it can distort the results of a total absorption precision standard cost system.

Total absorption standard costing is an excellent tool for use in inventory valuation, cost engineering, manufacturing management, operations evaluation, and cost management. It is also useful in defining cash flows. It is not intended to be a good tool for financial management of the enterprise.

All of the material, labor, and indirect or overhead costs of each operation must be collected and accumulated into the "bucket" as they are generated. As the product or part progresses from level-by-level through the operations of the system, the costs in the item's "bucket" or costed inventory record accumulate. The total standard and actual accumulated cost of the part or product is available at any point in its flow through the process by looking into the "bucket" or reading the inventory valuation of the work-in-process at that point.

The costs entered into the inventory record or "bucket" are in two forms. The standard costs are accumulated in the WIP inventory record in each case. The actual costs are also accumulated into the WIP record. Any differences or variances between these standard and actual costs define the affects of production performance, changes in methods or processes, problems in operations, skill variations, schedules, etc. These differences or variances provide management with control data feedback and a measure of the cost impact of deviations from plan.

This total absorption approach produces the kind of data base needed to resolve the conflicts generated by the often stated demand that the financial and cost management systems must interlock. This demand creates administrative obstacles to the implementation of precision engineered, total absorption costing systems.

The product engineering, manufacturing, and physical distribution operations need precise and completely detailed cost data based on a stable and measurable baseline or datum (i.e., time and fixed facilities). Enterprise management needs organizational or responsibility oriented accounting data which meets financial and tax management criteria. While these are conflicting requirements, today's computerized data base type accounting systems can respond to these conflicting requirements and accommodate management's need for both financial and operations cost data and feedback.

By computer manipulation of the common data base to serve a wide variety of management audiences, each can have their required reports and analyses. The same input data can be used for both financial management and engineering and operations costing. The differences in their reported results can usually be reconciled by thoughtful allocation of the computed variances to balance the precision costs against generally accepted financial controls.

The variances between the actual costs developed by these procedures and the distributed costs used for financial management of the enterprise should be applied as adjustments to the profit and loss statement. They should not be used to modify the computed cost of the product.

As pointed out in the discussion of level-by-level cost accumulation in Chapter 4, a time based standard cost approach provides a baseline or datum against which to measure the actual labor utilization and material costs, and define their variances from planned performance. Prior chapters also described the errors and problems which can result from the use of volume based burden and overhead "distribution" or "percentage" cost allocation procedures for charging indirect and/or period costs to the product (Figure 2.3). There is a better way!

Manufacturing costs are composed of many elements (Figure 5.1). If we extend the measured standard cost concept to include all of the direct, indirect, overhead and general and administrative expenses of the product or operation at each production step in the manufacturing or distribution process, we can cumulatively define the up-to-date cost of the product, part, or component at each level of the process (Figure 2.1).

This costing approach provides a basis for accumulating product and/or operation cost at any point in the manufacturing and distribution sequence. With this type of cost reporting, it is possible to:

1. Isolate the cost of individual parts, items, and operations.
2. Quantify the impact of methods or product design changes.
3. Define the cost impact of capital facility improvements.
4. Identify and promote high margin products and/or services.

Level-by-Level, Total Absorption, Standard Cost

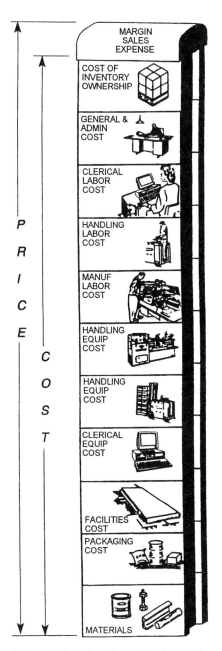

Figure 5.1 The elements of manufacturing costs.

The levels or control points where management authority can be applied through inventory control and cost accounting procedures can usually be identified by changes in the character, condition, or operational status of material or work-in-process moving through the system. Some of the situations which represent control points are:

1. A change in the condition or process status of a material or product in the production or distribution system which affects its shape, level of completion, work-in-process status, or storage posture.
2. A change in the shape or configuration of a material which affects its handling characteristics (i.e. cutting, coiling or uncoiling, boxing or crating, or palletizing).
3. A change in the identity, brand name, lot number, or part number of an item which affects its control status.
4. A change in the role or use of an item such as the conversion of raw materials to parts, or parts to sub-assemblies or finished products.
5. A change in the inventory level, or the sales or ownership status of an item from raw material to parts or parts to products; and from factory warehouse to wholesaler and/or retailer stocks.

Other breakpoints in the material control system include location changes such as moves within a factory, or from one building to another. Each provides a definable inventory transaction or record posting point which can also serve as a cost accumulation station. In modern systems, these record posting points are often automatic identification or bar code reporting points for material flow control and the tracking of the progress of the product through the process.

The materials management system which records the WIP inventory's status at each production level, and which reports the product's movements, storages, and changes in condition, can be used as the basis for the cost collection procedure. The material control transactions and the total absorption cost reporting can both be posted to the inventory record by either a manual or computer based procedure.

C. DEVELOPMENT AND APPLICATION OF THE "USE RATE" DATA BASE

While the only incorruptible basis for measuring performance is time, the building or floor space occupied by the operation also offers a potentially stable basis for comparative measurement of manufacturing and distribution operations. The space "use rate," as defined in Chapter 4, can be the vehicle for collection and application of indirect labor, burden, overhead, general and administrative expense, and other non-linear operating costs into the work station's "use rate."

As stated above, the net space used consists of the work stations and their supporting aisles and storage spaces. By excluding the areas occupied by toilets, cafeterias, lobbies, offices, and other non-productive spaces, the net operating space can represent the whole facility and its operating costs. The net space can then be the vehicle for the collection and application of indirect, overhead, G & A, and other non-linear operating costs into the work station's "use rate." Each square foot of the net operating space can serve as a cost module or "bucket" for the collection of these costs.

The work station's space "use rate" encompasses the total cost of operating the building or facility, including the cost of ownership, maintenance, taxes, insurance, interest, security, heating, lighting and ventilating expense, janitorial and sanitation costs, and any miscellaneous facility expenses. All of these expenses are accumulated annually. Their total is divided by the area of the net operating space and distributed to the net operating space as an annual cost per square foot.

In a manufacturer owned facility, the definition of this cost per square foot figure will require some supporting accounting procedures. If the building is rented or leased by the operator, the lease charge can be divided by the net operating area to define the basic space cost.

In most cases, accountants allocate building and equipment depreciation costs on the basis of the tax oriented depreciation schedules. As stated earlier, these are often rapid write-off depreciation schedules which present cost accounting problems because they do not generate a uniform annual expense.

The allowable tax life of the building or equipment also may not be the same as its true useful life. For example, if the tax law allows a depreciation life of 31.5 years, the building may actually serve the enterprise for a longer or shorter time. If the facility is a rack supported warehouse or another type of specialized structure, the building may be treated to the shorter equipment depreciation schedule for tax purposes and still be in service many years after it has been written off the books. In each of these situations, the apparent annual ownership cost of the building will be artificially defined by tax rules and financial considerations. This cost may vary with time and be incompatible with the true market rental value of the real estate.

The same definition problem exists in the ownership cost of manufacturing, materials handling, and storage equipment and computers. The equipment's annual cost must include its ownership cost, insurance, taxes, utilities, maintenance, and miscellaneous expenses. The IRS may allow five, seven, or ten years of tax related depreciation. The accountant may also be required to use the "Modified Accelerated Cost Recovery System's" (MACRS) rapid write-off technique. Conversely, the equipment may be made obsolete in less time than the tax rules allow, or continue in use for a much longer period. Definition of a realistic annual ownership cost is complicated by these issues.

To establish a defensible and stable ownership cost base which will also provide a replacement sinking fund, the author proposes an arbitrary, but logical technique for establishing a realistic ownership cost. This technique will be used in the development of the data base for the space and equipment "use rate." However, it will not be fully compatible with conventional financial accounting techniques.

There are two logical and defensible approaches to this problem. One is a market survey to select a typical or common lease or rental charge for similar space or equipment for use as our standard ownership cost. However, this approach is vulnerable to both local and national economic conditions, and to the truthfulness of real estate agents and equipment dealers.

A more logical, dependable, and defensible approach to the establishment of a "standard" ownership cost for real estate and equipment can be based on tax law and prime interest rates. The IRS directed (MACRS) tax life of the asset can be used as the basis for computation of a straight line depreciation charge. Then the Federal Reserve based prime interest rate can be applied for the tax life of the asset. This establishes a defensible ownership cost rate or rental charge as of the date of the purchase of the asset. This ownership or "use rate" charge should continue for the full actual life of the asset, regardless of its possible continuance beyond the tax life.

If the asset is disposed of early, the salvage income and unabsorbed ownership charges should be treated as an adjustment to enterprise profit—as miscellaneous income or expense. If the equipment or building is used longer than the IRS "life," the overabsorbed "use rate" will be treated as a profit or sinking fund to finance the replacement or upgrading of the facilities.

In the case of a planned short life of the asset, the shorter life span should be substituted for the IRS life allowance to compute a better and more accurate absorption of the ownership cost. This situation might typically occur with special machinery purchased for a short-term contract, or in the case of a temporary building which is to be replaced by a permanent structure. In either case, the product will be charged with the time use of facility "use rate" for each manufacturing operation; no more or less.

The computed "rental" charge provides a standard ownership cost which is included as a component of the work station's "use rate." This standard ownership cost contributes to the definition of a realistic product cost. It avoids the error of omitting facility ownership cost from the product's manufacturing cost when old and fully depreciated facilities and/or equipment are used in the process. This procedure also permits the accrual of funds for replacement of obsolete facilities and equipment.

The computed depreciation based ownership cost is added to the sum of the other accrued annual equipment or space expenses to produce a "use rate" base cost for each piece of equipment and each square foot of the net operating space. The product is charged for its actual time use of the facilities and equipment.

The total absorption standard cost concept also requires that general and administrative expenses and other manufacturing overhead costs be included in the work station's "use rate." The methods for accomplishing this and computating the complete work station "use rate" will be discussed later.

D. THE "COMMON DENOMINATOR" CONCEPT OF MATERIAL MANAGEMENT

The precision costing system can also be tied into the material handling system if that system is based on a "common denominator" handling module. The handling module can be used as the costing module. This is particularly appropriate if one recognizes that "the flow of material through the manufacturing and distribution system is the flow of funds through the enterprise" and "the control of material flow is the control of production and cash flow."

The "common denominator" concept provides for great flexibility in quantities moved. The handling modules are uniform and standardized. Pallets or tote boxes homogenize the handling system so that the shape, form, and quantity of the material being handled is "invisible" to the materials handling machinery.

Instead of the traditionally accepted modular control quantities of dozens, tens, or sets, the quantity of material in the handling unit varies with its size, shape and weight. The unit standard is defined as the standard quantity that fits into the tote box or onto the pallet or on the machine loading tool pallet. The procedure establishes the "common denominator" quantity for each item in the manufacturing or distribution system. This quantity is the "common denominator" materials handling unit or module. All moves and materials management transactions are in singles or multiples of these "common denominator" units or modules. In some cases, the traditional quantities can be coordinated with the "common denominator" capacity quantity.

The application of the modular "common denominator" concept is compatible with the principles of Just-in-Time operations and the desire of manufacturing management to reduce lot sizes. However, the smaller lot size or handling unit appears to be in conflict with some of the long standing fundamental principles of good materials management and materials handling.

In the past, it has been materials handling gospel to move the largest possible quantity as far as possible toward the next point of use before breaking down the handling unit or releasing it to the next operation or storage. This principle is not compatible with the reduced work-in-process inventory objectives of Just-in-Time operations or the flexible manufacturing systems' (FMS) ability to deal with lot sizes as small as one unit.

At the same time, materials handling practitioners have sought methods for simplifying and homogenizing the interfaces between the materials being handled and the materials handling equipment. The objective has been to establish a uniform materials handling system throughout an enterprise whenever practical.

These objectives have led to the development of homogeneous or "common denominator" pallet and unit load systems and modular standardized tote box programs which are compatible with both pallets and conveyors. This homogenization of the "common denominator" handling interface often permits the automation of materials handling operations.

By standardizing the module of movement and material control for each item as the standard quantity which fits into the tote box, on the pallet, or on the machine loading tool pallet, the cost accumulation becomes more easily managed. This procedure establishes a "common denominator" or standard quantity for each item in the manufacturing or distribution system which is also the "common denominator" materials handling unit or module. It can also be the standard quantity module for the production batch. All moves and/or materials management transactions will be in singles or multiples of these "common denominator" units (quantities) or modules. The cost system's data collection can also be based on these modules.

Such a system can be either automated or manual. In a Just-in-Time system, the moves should be in single modules or multiples of the "common denominator" module. Where possible and practical, the purchasing and production schedules can also be computed by using modules or multiples of "common denominator" quantities. Whenever possible, the finished goods sales quantity or lot (or set) should be based on a multiple of the "common denominator" modules.

From our point of view, a more important result is the "common denominator's" impact on the structure of the cost system. It provides a modular basis for establishing materials handling and storage cost standards which are fairly easy to accumulate and track. The "common denominator" module can also become the basis for defining the manufacturing cost standard for each operation. This could reduce the number of shop floor record keeping transactions and the computer can be programmed to compute the individual unit cost of the operation on the basis of the standard quantity module.

E. LEVEL-BY-LEVEL MANUFACTURING/WAREHOUSING COST APPLICATION

Because the primary objective of both the total absorption standard cost system and the time use of facility costing procedure is to facilitate collection and definition of fully accumulated costs for each item in production at each process level, this cost accumulation capability is important. It is useful in:

Level-by-Level, Total Absorption, Standard Cost

- The engineering and design of products to a target cost.
- The preparation of a preproduction cost estimate.
- The selection and costing of prior use components and parts for inclusion in new products.
- The valuation of work-in-process inventory.
- The analysis of manufacturing and distribution performance.
- The identification, definition, and evaluation of cost deviations from the plan.

Because the valuation of every item of material, work-in-process, and finished goods is fully costed at each level of control, the totally absorbed item cost allows the transfer of a costed inventory item from a work-in-process inventory status into the service parts role, or into the bill of materials for a new product, without any need for recalculation of its cost or value. Semi-finished parts or assembly components also carry their fully absorbed cost from all prior operations (materials, labor, overhead "use rate") into assembly or secondary operations. The secondary operations will add their costs when modifying the item into another part, component, or assembly.

These costing benefits, discussed in Chapter 4, together with the use of the part number's numerical bill of materials indent code to access the manufacturing bill of materials' matrix, provide a good basis for analysis of inventory dollar lockup, cash flow, and budget performance. This system also provides a means for identifying and valuing work-in-process inventory bulges and desirable "surge" inventory holding points.

From the engineering point of view, the totally absorbed standard cost helps preproduction estimating of new products or operations and improves the ability to accurately design to a cost target. In each case, the engineer is assured that the value shown in the costed work-in-process inventory will be a valid basis for predicting the whole cost of using the item in the new design. No extraneous or below the line costs are hidden for later surprises. Figure 5.2 shows the structure of the totally absorbed, level-by-level cost computation.

F. LEVEL-BY-LEVEL INVENTORY VALUATION AND PRODUCT COSTING

In combination with the application of matrix type numerical bills of material, and level-by-level material management procedures, precision cost accounting permits management to isolate and control detailed elements of production and distribution costs.

- It allows the prediction of product and parts costs.
- It provides a dependable basis for isolating and costing individual operating procedures and processes.
- It permits level-by-level analysis of operating costs on an item, operation, department, process, and customer basis.

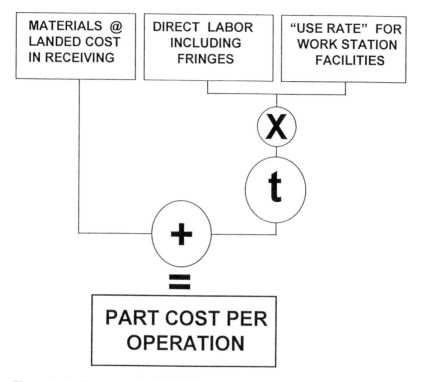

Figure 5.2 The concept of Part Cost computation based on the "use rate" approach to period and indirect cost absorption. This method is independent of production volume.

- It provides data for reliable responsibility accounting.
- It provides fully absorbed, and volume independent, part, component, and product cost information.

Multiproduct companies with complex products that contain common components and varying levels of market access can use these procedures as the basis for estimating product and service parts costs, evaluating price performance, and predicting margins.

Good cost information is the most elusive of management data, and the most critical to the manufacturing engineering, product design, and marketing decision processes. Total absorption standard costing and level-by-level material control provide top executives with the tools and data required to beat competition.

These techniques are applicable in either an MRP or a JIT environment, or in a combination of these materials management environments.

6

The Elements of Cost in Manufacturing and Warehousing

A. LABOR

To develop a sound cost system, it is first essential to clearly define the elements of cost. Let us first take a closer look at labor. Labor cost is not a simple wage per hour or salary per year. Labor costs are a composite of wages and many other supplementary or "fringe" elements.

Federal and state labor laws establish a minimum wage floor. Added to this are such legally required costs as the employer's share of social security taxes, worker's compensation insurance, and unemployment compensation taxes.

Union contracts also have an effect. They specify wages, hours, bonuses, vacation and holiday pay, health insurance, retirement plans, educational programs; and in the future, child care support and child care facilities. The total of these expense items makes up the prime cost of labor. Wages and salaries are only the base of the labor cost computation.

Another required and variable labor expense is the "for nothing" cost. Most of the "for nothing" cost is caused by the overtime labor premium. It is the cost of work scheduling. The worker earns the basic eight hour pay rate at the usual production rate. The premium is the result of a night shift or overtime schedule which generates a wage premium, but does not result in an increase in the volume of work performed in each night or overtime hour. The work done is the same as would be produced on the day shift during a regular eight-hour day or forty-hour week at the basic wage rate plus fringes.

The "for nothing" wage premium includes the 50% bonus paid on a regular wage for working hours in excess of eight hours a day or forty hours a week, the double time bonus (100%) on Sundays and holidays, and the 10% or 15% night shift bonus. The "for nothing" cost includes the base pay plus additional social security and worker's compensation taxes, and unemployment insurance. The night shift or overtime schedule increases the price paid for the labor but the wage premium is paid "for nothing." There is no increase in productivity.

If the operation is based on regularly scheduled overtime or night shift operations, the overtime or shift bonus should be treated as an adjustment to the basic labor rate and be averaged over the standard pay period (i.e., weekly etc.). In such cases, it should not be treated as a "for nothing" cost but as an evenly distributed payroll adjustment.

If the overtime schedule is irregular or variable, the resulting "for nothing" labor cost is a variance and a cost of scheduling. It should therefore be charged to the enterprise management as a variable indirect overhead expense and an adjustment to profit.

Since "for nothing" cost is an indirect manufacturing expense, and is not a direct production labor cost, it should be fed into the product cost via the "use rate." The "for nothing" cost should only be charged directly to the product if it is a payroll adjustment to distribute and absorb regularly scheduled overtime or night shift operations, or if the overtime or night shift schedule is definably customer driven and the schedule is at the customer's request. In most cases, the "for nothing" cost should still be handled in the same way as any other overtime premium. However, the accounting department may want to add a surcharge on the invoice so that the "for nothing" cost can be billed to the responsible customer.

Another variable labor cost is generated by the use of incentive wage programs, bonuses, and profit sharing plans. It is usually assumed that productivity increases as a result of the higher take home pay that these programs give to the employee. Although this is usually the case, social or peer pressure sometimes limits the effectiveness of factory incentive plans. In such cases, the productivity improvement may not be linear with the labor cost variance.

Management bonuses and profit sharing plans for manufacturing supervision and supporting employees are intended to have the same productivity improvement results. However, they are not as easily related to the performance of the actual production operation.

From a cost accounting point of view, there are at least two approaches to the introduction of the cost of incentives into the product cost. The average annual cost of the incentive program can be treated as an item of factory overhead and distributed to the product's cost and inventory value via the "use rate" along with other variable overhead expenses. In an incentive operation, the

The Elements of Cost

increased productivity will result in a negative production time variance as demonstrated by an increase in the output per hour. This will normally result in a reduced unit labor cost and a lower "use rate" time charge with a negative cost variance or cost reduction for the operation.

This negative product cost variance will be partially offset by inclusion of the incentive pay cost in the "use rate" as a part of the overhead expense. Thus, the labor and "use rate" costs of the product and the operation will decline, but the overall payroll cost will be increased by the incentive. As stated above, this payroll variance is absorbed into the product cost as an element of the "use rate."

The incentive wage can also be treated as an hourly payroll variance for each individual employee or as a total payroll variance on the enterprise as a whole. On an employee basis, the incentive can be charged directly to the operation and the product. This approach has the advantage of pinpointing the impact of incentives on individuals, but it also complicates the standard cost procedure. From a personnel management point of view, the individual effect of the incentive is important and can be tracked in the payroll account. From a cost accounting and estimating point of view, the overall effect of the incentive program is a more critical issue and may affect standards. In any case, the incentive wage cost and its variable productivity impact must be introduced into the product cost computation. In a computer based system, management can have it both ways from the same source data.

In the proposed total absorption, time use of facility system approach, the cost of the incentive will be treated as an element of manufacturing overhead and will be included in the "use rate." It will find its way into the product cost parallel to the individual employee's wage for each operation. The "for nothing" cost will also be introduced into the product cost via the "use rate."

In the financial records, both the "for nothing" cost and the incentive wage payments produce a variance from the predicted standard payroll cost of the operation as a whole. This effect is recognized in budget analysis procedures and in the profit and loss statement. These costs are also captured by the "use rate" and included in the costing of the product.

Therefore, the labor cost (L) in manufacturing and warehousing consists of:

- Basic employee wages or salaries (hourly rate) = W
- Legally required payroll taxes and insurances = T
- Management/union sponsored fringe benefits = F
- "For nothing" costs based on schedules = N
- Incentive pay and bonuses for performance = B

In a manufacturing or warehousing operation there are also several different categories of labor cost. Each requires a separate accounting treatment. The most obvious of these categories are:

1. Direct labor (L_D), which changes the product's condition, status, shape, packaging, appearance, or function.
2. Indirect labor (L_I), which logistically supports, handles, moves, inspects, controls, sets up, or otherwise affects the product's flow through the manufacturing process without changing the product itself.
3. Burden labor (L_B), which provides supporting supervision, clerical and administrative personnel, labor relations, maintenance, safety, janitorial, and security activities without direct product contact.
4. Manufacturing management labor (L_M), which includes the Vice President of Manufacturing or Operations Executive and their supporting staffs, the product design and plant engineering staffs, information system, production, and inventory control personnel, etc. These people are also a part of the burden labor pool and their cost should find its way into the product cost via the "use rate."

In most manufacturing operations, direct labor is fairly easy to identify and define. When the operation is running in the factory, direct labor time can be measured by a variety of work measurement techniques (stopwatch studies, work sampling, video, predetermined time standards, historical records, etc.) and labor reporting systems. Cost can be calculated from accounting records. Direct labor time can be captured through the worker's normal time reporting parallel to the production operations' tracking and machine time reporting procedures. Much of the indirect labor can also be defined in the same manner.

In planning a new operation, estimates of direct labor times, and some of the indirect labor times, are usually based on planned machine running times, planned supporting operation times, and/or a detailed work station analysis using such predetermined time standards as MTM or MOST, or comparable times for existing or past operations.

When a standard time (or predetermined time system) and standard cost system is in place, the direct labor cost of each operation can usually be estimated by the manufacturing engineer. The total predicted standard time for each part or product's manufacturing operations can then be computed and entered on the operations sheet or routing. After the operations are running on the shop floor, the times can be verified by work measurement and corrected to provide a valid work standard. Schedules and standard costs can also be computed from the standard times and operations times can be listed for the part or product on the operations sheet or routing.

Actual operation time variances and labor cost irregularities can be captured from the employee time reporting procedure and/or machine running times. In many modern systems, bar code wanding and/or other automatic identification techniques are used as data and time collection media. The ac-

The Elements of Cost

tual cost can be computed from the time reporting system. In systems which lack a standard time or cost base, the actual recorded operation time can be used to calculate the actual cost, but there will be no standard against which to evaluate performance. When there is a standard time and standard cost system, the actual time and cost can be compared to the standards to compute the variances and trigger any required corrections in the operations or standards.

"Touch" type indirect labor will normally vary with the rate of production and run parallel to direct labor. "Touch" indirect labor would include machine set-up people, inspectors, materials handlers, storekeepers, warehousemen, and the preventive maintenance staff. However, their relationship to the operation is not always uniform or constant. For example, the inspectors may only sample the work and not handle every piece, and materials handlers may handle the material in large batches.

However, many "touch" labor functions can be defined as standard operations in a labor standard and can be applied to the flow of work through the process sheet. They often have a standardized and/or predictable, but often non-linear, relationship to a production work station.

From an accounting and cost point of view, these workers can sometimes be treated in the same manner as the direct "touch" labor. In such cases, it appears to be practical to directly apply these labor elements to the production of the product through an operations based standard cost procedure. However, this is not usually the case. Therefore, this indirect labor (L_I) should normally be treated as an overhead item and flow to the product cost through the "use rate" procedure.

Burden labor presents a different problem. The burden labor is a period expense which seldom varies with manufacturing activity. This would include "no touch" labor such as engineering, clerical, scheduling, supervision, security, janitorial, and manufacturing management activities. This "no touch" burden labor expense must be treated as a general overhead item. This requires its application to the product cost through the same distribution procedure used to deal with other period or indirect expenses of the business. These period labor costs can be equitably allocated to the various operations in a time-based manner through the "use rate."

In some cases, product design engineering is treated as a separate production related activity excluded from the burden labor and expense category. In such instances, the design engineering function can be treated as a distinct work station or process, and its costs can be collected on a product costing basis. This type of operation can include preliminary design, detailing, model shop operations, product testing, and prototype production. The costing procedures are the same as used in the manufacturing process, and the design activity related to each product is treated as a process unto itself.

With the exception of the "no touch" indirect or burden labor expense and the manufacturing management labor, both of which will be treated as elements of overhead cost, the cost of labor can and should be applied to the product on an operation-by-operation, time use basis. As stated earlier, the only positive and invariable element of measurement in production is time. By summarizing labor cost elements into time cost units, and by applying labor cost to the product on a time basis at each step in the process and for each piece of the product, the accuracy of the labor element of production cost can usually be assured.

B. MATERIALS

Though the definition of material cost appears to be simple and obvious, this is not usually the case! There are several classes of manufacturing materials which must all be tracked and captured by the cost accounting system. These include direct materials, indirect and/or contingent materials, burden materials, consumable tooling, power, fuel, and office and sanitary supplies.

Defining these material costs requires the use of a combination of purchase prices, work-in-process inventory valuations, and consumption analysis. The accountant can choose one of several inventory valuation techniques for purchased material and the parts and components produced in the plant. The most commonly used methods are:

- FIFO—First-in/First-out: In this case the inventory is underpriced on a rising market and overpriced on a falling market.
- LIFO—Last-in/First-out: This method overprices on a rising market and underprices on a falling market.
- AVERAGE—In this case the trends are flattened but the value continues to lead or lag the market.
- STANDARD—This method is the most commonly used in modern cost accounting practice. A period cost is established. If the variance between actual and standard cost is random or cancels out for the period, the "standard" is correct. If it is "skewed" and shows a cost trend, the material cost standard must be adjusted (Figure 2.5).

For the purposes of this discussion, the standard direct material quantity will be defined as the design standard material required for the planned method of manufacture. This will be M_D in the calculations. But the planned standard material is not always actually used. Material variances develop from:

- Schedule or quantity changes which result in different consumption rates, vis-à-vis the standard.
- "Nest and gain" effect on material usage.

The Elements of Cost

- Scrap, and/or "chip" losses, and shrinkage.
- Learning curve effect on scrap loss.

Direct materials are those which become a part of the product and include the scrap which is generated by chips and errors in the fabrication process. Direct materials can usually be classified as raw materials or purchased components. The materials and purchased components which are used in the product are usually defined by the design specifications, the engineered standard quantity for each part, and the manufacturing bill of materials.

In most manufacturing operations, the required standard material quantity is greater than the net material remaining in the part when it is completed. If the part or component is machined, cast, molded, stamped, forged, cut out, or otherwise formed, it is common to have chips, drop off, risers, flash, or other surplus material removed from the part during the manufacturing process. The surplus can often be recycled, which will reduce the total net cost of direct materials to below the projected standard material cost.

However, the standard cost of the raw material is normally calculated on the basis of the standard unit of input material used in the process. Therefore, the cost recovery from the sale of the recycled scrap or surplus material is a negative cost variance which can be applied to the total purchase quantity and the standard cost of the material, or it can be treated as an adjustment to profit. This book's approach treats the surplus or scrap sale revenue as a miscellaneous income item and an adjustment to profit. This causes the total material input to the process to become the basis for establishing the standard material cost.

This is only one of the variables involved in the costing of the direct materials. The materials' variables also include the following:

1. Nest and Gain

One of the most important, and most frequently overlooked factors in material requirements planning is the "nest and gain" factor (F_{ng}). This is the material effect of scheduling and lot quantity. This variance is usually generated by the different methods and sequences of production which are a function of lot size and schedule. These "nest and gain" effects are most noticeable in bar feed operations and in such "flat" part production as sheet, plate, cloth, paper, and leather pattern layout operations.

In most manufacturing operations, the "nest and gain" impact is a function of the product line and the work scheduling. If the product mix and work load are fairly consistent, the "nest and gain" practices will stabilize and become part of the normal materials scheduling of the operation. This would be likely in clothing production (pattern layout and cutting), screw machine operations (fasteners and small parts), and other stable manufacturing situations. In such cases, the adjusted material usage can be defined as an average for each material class, and standard "nest and gain" factors can be computed. These factors can be used in the calculation of material requirements standards for material planning and control, and in computing standard costs.

If scheduling irregularity prevents the prediction of the "nest and gain" factor from historical material usage rates, the impact of "nest and gain" is treated as a material purchase cost variance. Its impact is handled in the same manner as scrap loss, chip, and shrinkage. The cost or savings variance is captured in the material cost account rather than charging it to a specific product or operation. The "nest and gain" factor is the material impact of scheduling and should be charged to the manufacturing cost through an adjustment to the overall material cost. Alternatively, it can be treated as a miscellaneous expense reduction or a material cost reduction in the profit and loss statement.

2. Scrap and Chip

The modern trend is to try to achieve "Zero Defects" (ZD) through "Total Quality Management" (TQM) and to minimize scrap losses. But nobody is perfect! There will always be some errors and scrapped material in a manufacturing operation.

In modern manufacturing philosophy, the objectives of "zero defects" and "Total Quality Management" demand the elimination of scrap from the system through elimination of errors before they occur. When computer controlled machinery and very tight management techniques are used, these objectives can often be achieved with some success. However, many manufacturing operations are still human dominated and subject to many random variables.

In the past, scrap allowances were predicted on a percentage basis to assure adequate in-process inventory and to avoid shortage caused downstream shutdowns and delays. This often resulted in excessive work-in-process (WIP) inventory and surplus materials and parts. The trend toward Just-in-Time (JIT) manufacturing scheduling has focused attention on this problem, and the need for better error and scrap prediction has become evident. The need for error prediction, and the inventory and cost implications of errors continues to be present, especially in labor intensive operations. The learning curve is sometimes applicable in these situations, but errors will continue to occur.

The theories of statistical analysis and the laws of probability have long been applied to this error prediction problem. Where adequate data is available, statistical analysis can usually predict scrap losses with an acceptable degree of accuracy. Scrap allowances (M_S) can then be realistically included in the material standards.

The old method of applying percentage factors is not acceptable in today's manufacturing environment because it inflates the absolute scrap allowances as the lot quantities increase, and it generates excess safety stock inventories. This

The Elements of Cost

is contrary to the trend toward lower work-in-process inventories and Just-in-Time management techniques.

In the absence of statistical analysis procedures, a reasonable, but approximate method for predicting scrap allowances uses the square root of the lot quantity as the basis for predicting the required extra pieces. This technique limits the growth of surpluses and flattens the scrap allowance curve. This approach will be discussed later.

In any case, the predicted or actual scrap allowance should be charged to the cost of the product as a computed part of the material cost. Any variances in predicted scrap loss or inventory shrinkage costs are captured on a product basis in the material inventory record. The cost of the scrap variance should be treated as a material cost variance. This variance will then be captured in the profit and loss statement.

"Chip" material is that which is removed from the standard material (M_D) as a part of its conversion into the "net shape" of the part. Cost recovery by the sale of "chip" is treated as a miscellaneous income in the profit and loss account, not as a material variance.

3. Indirect Material

Indirect material is another category of manufacturing material cost that must be assigned by the costing system. It is the indirect or contingent material (M_{ic}) which is consumed in production but does not become a part of the product. This material includes items such as abrasives, lubricants, foundry sand, consumable tooling, coolants, catalysts, fuels, heat treating chemicals, protective clothing, work-in-process packaging and product protection materials. Electricity and other utilities used to operate the manufacturing process can also be treated as indirect materials.

In some cases these materials can be attributed to a specific work station. Their cost can then be allocated to that work place as a part of the "use rate." In most cases, however, these indirect materials should be treated as a burden to the manufacturing operation. They will be considered along with burden materials.

4. Burden Materials

Burden materials (M_B) can best be defined as those materials which support the manufacturing operation but are not actually used in the manufacturing process. Examples of these are sanitary supplies, maintenance materials, forms and documents, and first aid supplies. In addition, the fuel and electricity used in the heating, lighting, and air conditioning of the plant, and the water supply can be considered as either space cost or burden materials. Because most of these items are purchased on a continuing basis, they constitute a period cost.

Burden materials are not made a part of the product, and their consumption is not normally linearly related to the production process. The cost of these materials should be distributed to the product cost through the "use rate" and the time use of facilities along with other indirect expenses. This procedure will be discussed later.

C. INDIRECT OPERATING EXPENSES

All cost accounting systems and "generally accepted accounting practices" recognize and try to capture and quantify the primary elements of indirect business expense. These include all indirect labor, materials, and expenses of the enterprise generally referred to as corporate "overhead" and "general and administrative expense."

To reduce clerical workload and simplify the development of management criteria in the past, "overhead" was frequently summarized as the total of all period and indirect expenses. It was then applied to the product cost by dividing the total indirect expense by the number of product units produced or the number of labor or machine hours utilized in production. To apply overhead to individual products or processes, this quotient was then often converted into a multiple or percentage of either labor or material cost.

In some cases, this computation also included an allowance for general and administrative expense. G & A includes the president's salary, other top executives' salaries, and the cost of their offices and staff. In most cases, however, the G & A expense has been treated separately as a "below the line" expense and has not usually been included in the manufacturing overhead or assigned to product unit cost. We will consider a change in that practice.

These past approaches inaccurately relate the indirect and/or period costs to the product cost. They do not correctly define the relationship between indirect and/or period expenses and the direct manufacturing cost. They do not focus the proper share of the overhead cost on the specific product, process, operation, or department.

"Generally accepted accounting practice" is also inconsistent in the handling of "overhead" expenses. Accountants differ in their definitions of factory burden, factory overhead, burden and/or overhead materials, indirect and burden labor, and general and administrative expenses. They also differ in their methods for introducing indirect expenses into the manufacturing cost accounting system and the absorption of these costs into the product cost.

In most cases, accountants use some form of distribution accounting as discussed earlier. The effect of these distribution techniques is to imply an equal or labor based allocation of cost to all items in production and to vary the allocation in an inverse ratio with production volume. This is not a true picture

The Elements of Cost

of the behavior of manufacturing costs. There are better and more accurate methods!

For the purposes of this text, let us define the indirect operating cost elements of "overhead," "burden," and other indirect manufacturing expense. General and administrative expense will be treated as a separate element of cost.

Burden in precision costing consists of those indirect and period costs directly related to the manufacturing, warehousing, and distribution activities at the manufacturing or operating level. Burden costs include some indirect labor, all burden labor, burden materials, maintenance costs, and the facilities' operating expenses required to support production operations without becoming a part of the product.

Overhead consists of those elements of labor, materials, and facilities expense required to operate the manufacturing or warehousing enterprise but do not become directly involved in the manufacturing or production operation. These include the enterprise's administrative functions and their offices, production and inventory control activities, security, janitor services, human resources (personnel) management, product engineering and research, supplies, and the cost of owning and operating the required buildings, facilities, and equipment.

The primary elements of manufacturing burden and/or overhead are as follows:

Indirect labor (L_I) is the labor which is a part of the production system, but does not change the product. It is often "touch" labor, but cannot usually be allocated to a specific operation. This category of labor includes the general inspection staff, general materials handling and store room labor, receiving and shipping labor, maintenance staff, general set-up personnel, tool makers, and tool maintenance people. The cost of this labor is allocated to the operation and product through the "use rate" procedure.

Burden labor (L_B) is the labor which supports manufacturing but does not "touch" the product. This labor category would include the factory management, department managers, supervisors, production and inventory control staff, computer operators, manufacturing and industrial engineers, labor relations and personnel management staff, safety and environmental management personnel, nurses and doctors, plant engineering and maintenance staff, general factory office staff, janitors, fire fighters, and security guards. The cost of this labor is charged to the product and operation through the "use rate" procedure.

Indirect materials (M_{ic}), as discussed earlier, are those materials used in the process of manufacturing that do not become a part of the product. In some cases they are used proportionally with the direct materials and can be charged into the product cost in the same way as direct materials. In other situations, they may be limited to a single department or process and can be allocated to the burden or overhead of that function. However, most of the indirect or contingent materials are used in a non-linear relation to the direct materials. Some of these are abrasives, foundry sand, plating electrolyte, catalysts, coolants, consumable tools, protective clothing, etc. These materials must be treated as burden expense items and charged to manufacturing and product cost through the "use rate."

Burden materials (M_B) are those items consumed in the operation of the enterprise or manufacturing facility, but are not directly related to the manufacturing or warehousing operation. As stated earlier, examples of these materials are floor sweep, sanitary supplies, stationery, building maintenance supplies, etc. The cost of these materials is also distributed through the "use rate"

Facilities expense consists of the ownership cost and the operating expense of the buildings and equipment used in the manufacturing and/or warehousing operation. These costs include depreciation or rent, utilities, insurance, property taxes, licensing fees, maintenance, and security systems. These expenses are the primary or core element of the "use rate."

D. SPACE AND EQUIPMENT COST

A major accounting issue is the method of applying the cost of owning and operating capital facilities to the cost of the product. This investment includes the land and buildings which house the operation, their attached building support systems (i.e. fire defense, utilities, telephone, etc.), environmental equipment (heating, ventilating and air conditioning [HVAC], lighting, etc.), the manufacturing and materials handling equipment used in production, the office, factory, and warehouse furniture, and computers. The cost of owning and using these capital facilities must be included in the cost of manufacturing the product.

There are almost as many methods for capital cost allocation as accountants who design them. "Generally accepted accounting practice" usually includes these costs in the general distribution of indirect expense to the product through labor or machine hours. A more equitable allocation procedure does exist.

Equipment cost and its allocation to the product is a good place to begin the discussion of burden expenses. Each piece of manufacturing equipment and machinery, and each piece of materials handling and storage equipment has a definable purchase or investment value and usually, a predictable service life. However, the Internal Revenue Service (IRS) and other countries' taxing agencies set official depreciation lives for each class of equipment and for buildings. In most cases, the manufacturing equipment's true service life is different from the depreciation life set by the IRS. This is true because the owner either uses the equipment longer than the IRS allowable life or advances in technology make it obsolete and in need of replacement before the tax life expires. As stated earlier, many financial managers also use rapid write-off depreciation for tax benefits.

These factors create problems in establishing a standard cost factor for the allocation of equipment depreciation and ownership expense to the product. The same problem applies to buildings because they usually last longer than the IRS's allowable depreciation life, or the enterprise's growth and change make them obsolete earlier.

As in the case of material's scrap, the salvage value of an obsolete building or piece of equipment should be treated as miscellaneous income and as a variance in the profit and loss statement.

The "precision costing" philosophy is based on the assumption of a continuous and stable ownership cost for capital equipment and buildings, and not on the assignment of a specific and limited investment service life. By using a straight line depreciation schedule based on either the IRS allowed life or the realistically predicted service life of the asset, we can establish a fixed ownership and use charge or rental rate. The base value for calculating the "use rate" is the purchase price of the equipment or building. A confirmable or actual market lease rate can be substituted for the computed rental charge in some cases.

However, depreciation is not the only cost involved in building and equipment ownership or operating expense. Other cost factors include the interest on the financial investment in the asset, and the cost of insurance, taxes, maintenance, power, and utilities. All of these expenses are a part of the cost of owning the asset and the space the operation or equipment occupies. All of these items add up to the annual cost of the facility or the individual work station.

But some of these facilities and equipment are not directly used or involved in the manufacturing or warehousing operation. They are the offices, toilets, entries, stairwells, elevators, heating and air conditioning facilities, firefighting equipment, security systems, fences, etc. which are a part of the building, but not directly involved in the process. Nonetheless, these facilities must be paid for and their ownership cost must be captured for inclusion in the valuation of the inventory and later recovery in the sale of the product.

To deal with this problem, we define the "net operating space" occupied by the manufacturing operation and warehouse and their supporting aisles. Using this space as the cost distribution base, we divide the computed annual cost of owning the whole building and its attached building related equipment by the area of the net operating space. This generates a fully absorbed annual facility rental rate for each square foot of operating space. This process absorbs the total annual cost of the building and its attachments into the space charge or "use rate" for the operating space.

Continuing this philosophy, we divide this annual "use rate" by the scheduled hours of operation (i.e., 2080) to get an hourly building "use rate" per square foot, per hour. By defining the space occupied by an operation, work station, or department, and multiplying the area by the hourly "use rate," we develop an hourly space charge for the operation. This is the first step in the development of the space "use rate" as the vehicle for allocating indirect expenses to the operation or product.

A similar procedure can be used to develop a "use rate" for machinery and manufacturing materials handling and warehousing equipment. In this case, the straight line depreciation charge should be based on either the IRS allowance or the expected service life of the equipment, whichever is least. This will assure a fair annual rental rate and the generation of a fund for upgrading and replacing the equipment as technology advances and it becomes obsolete. The annual charges ("use rate") should also include allowances for interest, utilities, insurance, maintenance, and the ownership of spare parts inventories. The annual cost of the equipment should again be divided by the scheduled use time for the machinery to develop the hourly based "use rate." If the equipment is only used part of the time (as in the case of a jig borer or a broach), the scheduled use time would of course be less than the normal 2080 hours per shift year. The "use rate" would be calculated accordingly.

Thus, the basic hourly rental rate or "use rate" for a work station would be the sum of its equipment "use rate" plus the product of the assigned work station's area times the facility's space "use rate" charge per square foot of space. Additional costs will be added later to make the work station's "use rate" the vehicle for collection of all indirect operating costs.

E. GENERAL AND ADMINISTRATIVE EXPENSE (G & A)

G & A is usually defined to include the salaries, office expenses, furniture, equipment, space, and supporting supplies used by the top management and their staffs. This includes the board of directors, the chief executive and all of the other corporate officers of the enterprise and their supporting staffs in the financial, accounting, computer services, marketing, public relations, and legal counsel functions (Figure 6.1).

As mentioned above, general and administrative expense is generally treated as a "below the line" cost by most generally accepted accounting practices. In this attempt to build a level-by-level, total absorption standard cost system, we offer an optional approach which can fully absorb G & A into the product cost at each stage of its manufacture. This approach allows the full absorption costing and evaluation of work-in-process and finished goods inventories. It includes G & A as a part of the indirect cost of manufacturing the product. To do so requires a new theory as the basis for G & A distribution to the product at the factory and warehouse level.

The basis for the proposed G & A distribution and allocation theory is the arbitrary assumption that the whole purpose of the executive suite and its supporting organization is to "manage the invested capital for the benefit of the stockholders" of the enterprise.

The Elements of Cost

Figure 6.1 The general and administrative cost block.

It is also assumed that invested capital can be defined as the total of the purchase cost (not the depreciated value) of the equipment, buildings, and land used by the business, plus the working capital, securities, cash, and inventories owned by the enterprise.

Although this is a true definition of the "invested capital," it is probably impractical to build a cost standard upon a base which includes such variable elements as securities, cash, and inventories. Therefore, our definition is arbitrarily restricted to the initial investment in such hard assets or fixed assets as capital equipment and real estate or, in the case of leased facilities, their market value at the beginning of the lease.

On the basis of these assumptions, the G & A expenses can then be said to be appropriately allocated to the operation through their relationship to the total invested capital. The value of "invested capital" can be a "standard" based on the original fixed capital investment in hard assets and possibly the total average value of working capital, cash, and securities. However, as stated in the previous paragraph, in order to stabilize and simplify the calculations, we define invested capital as only the purchase cost of land, buildings, and production and materials handling equipment.

Using this concept, we divide the annual G & A expense by the total of the capital invested in the hard assets of the enterprise to arrive at the annual dollar value of G & A expense per dollar of invested capital. This value is then divided by the scheduled hours of operation to produce an hourly G & A charge per dollar of invested capital.

This factor will be used as a component of cost in the later development of the work station's complete "use rate." This is accomplished by multiplying the hourly G & A charge times the dollars of invested capital in the work station as represented by the purchase cost of its equipment and space.

A consideration in dealing with the G & A expense's allocation to the product cost is the fact that many companies consolidate the administrative services required for several divisions and/or subsidiaries into central offices and headquarters operations. In such situations, the allocation of G & A must first be distributed to the division or subsidiary level and then to the plant level before the proper share of G & A can be defined for allocation to the product through a "use rate" procedure.

Sales organizations are usually treated as a separate operational element of the enterprise. Their expenses are a part of selling expense in a "below the line" posture. They are not usually a consideration in the definition of manufacturing cost.

F. THE COST COLLECTION AND BUILD-UP PROCEDURE

In the past, one of the primary obstacles in the development of precision cost accounting procedures has been the high clerical cost of data collection, manipulation, and analysis. We have long had the ability to record materials, work-in-process, and finished goods inventories on a periodic basis. We have also had the ability to record the workers' hours of attendance and the labor time on a task. The limiting clerical workload has been in the tracking of product through the factory, the manipulation of burden and overhead data, and the progressive analysis of production costs. Manual systems have also suffered from clerical error problems.

Manual methods, and even the linear methods based on punched card systems, have usually been capable of manipulation and analysis of the data, but they generated long data analysis and reporting lead times and high clerical cost. The information lag varied from several days to a month or more in many cases. This pattern is not responsive enough for the management of modern high technology operations.

Modern computer based information systems, and the development of data base software, automatic identification, data transmission, and computer networks, have simplified and accelerated the data collection process, minimized clerical error, and linked the data collection systems directly into the computer. This technology has also increased the speed and adequacy of data manipulation and analysis. Today's technology can offer near real time production progress reporting plus near real time cost and data feedback to the manufacturing operation's management and the enterprise's management reporting systems.

The speed of state-of-the-art computers permits manipulation of detailed data bases and presentation of a variety of output data for on-line manufacturing controls, machine operations, report preparation, and computer terminal presentations. The whole manufacturing operation can be made transparent by putting a computer terminal on each executive's desk so that he can call up selected data and answer management questions on an on-line basis. This capability permits the concurrent development of precision manufacturing cost data and manufacturing operations information. This makes real time manufacturing control feasible.

Because the control of the material flowing through a manufacturing enterprise is the control of the enterprise, all costs must flow into the level-by-level valuation of the work-in-process and finished goods inventories. All revenues flow from the sale of the finished goods. It is, therefore, obvious that collection of labor, material, burden, and production status data in a timely manner requires a detailed progress reporting system.

From a bottom-up perspective then, the elements of actual cost can all be collected into the data base via a bar code recording of each transaction. By using a system of part number, document number, and employee number coding and identification, all transactions can be related to the data base. This data base will contain the standard data on all the elements of cost, and the formulas to compute the actual cost variances from standard. The reports or computer responses generated will help adjust the deviations from planned operations back to standard. The computer will also produce standard and actual product budgets on a timely and repetitive basis for management monitoring of the operation. These same source data can also be rearranged into the input for financial control procedures and general accounting activities.

G. THE OPERATING AND PRODUCT COST STATEMENT

To give management a product based performance evaluation tool, the product cost statement must include all of the cost factors discussed in this chapter. The product cost statement should include all of the labor, material, burden, overhead, and general and administrative costs. The product cost statement in its final form establishes the factory cost of sales or the landed cost of sales. The landed cost of sales can be defined at the manufacturer's shipping dock if an FOB shipper and the customer pays the freight, or, if FOB destination with transportation cost included, at the customer's receiving dock.

In the sample computations presented in this book, we will omit the transportation cost from the factory or warehouse shipping dock to the customer when computing the "landed cost of sales." We will define the "landing" as the factory or warehouse shipping dock. The reason for this omission is that transportation costs will vary with each shipment, mode of transport, destination, and freight classification. The cost of transportation is a whole subject unto itself and should be treated separately as a part of traffic management.

The "landed cost of sales" calculation moves a step beyond the "factory cost of sales." It provides a basis for many key marketing decisions. The landed cost is the customer's view of the competitive value of the product. It can be the determining factor in the decision to market a product in a particular area or territory. It is a critical issue in the competitive analysis and selection of distribution methods and in the definition of logistics costs.

H. THE COST STRUCTURE

Before computers, it was impractical, and often impossible to develop data intensive, targeted manufacturing cost reports. Pinpointing the cost of manufacturing in sufficient detail to compute a precise product or process cost history or cost estimate was often difficult. As a result, most accountants used ratios and percentages based on generally accepted distribution type accounting practices.

In modern, capital intensive global manufacturing operations, the competitive environment prohibits such approximate methods for cost determination. Computers, data base software, and automatic identification techniques, in conjunction with detailed cost analysis and segmentation of cost elements, allow the accountant to collect information into a detailed data base. The computer can then manipulate the data base into customized and user or function oriented reports on a near real time basis. Survival and growth increasingly depend upon the accuracy of today's customized, near real time, user/function targeted cost reports.

I. SUMMARY OF THE COST PHILOSOPHY

To set the stage for some of the concepts presented in this book, it is appropriate to restate and summarize some parts of the author's philosophy of management.

The "use rate" and "time use of facility" philosophy presented in this book is an attempt to achieve a more accurate and detailed production and product cost definition. The following are some of the thoughts or concepts upon which the theme of this book is based.

- Price is usually independent of cost and is set by the market place and the competition.
- Cost is a function of management's performance and the economic environment.
- Profit is the difference between cost and price—the scoreboard of top management performance.

The Elements of Cost

- The flow of materials and products through an enterprise is the flow of funds and, therefore, the level-by-level control of materials flow is management control of the enterprise.
- Time is the only reliable, measurable, predictable, and incorruptible element in the cost equation and should be used as the basis for defining and managing all other elements of the operation.
- All costs should be allocated to the product and included in the inventory valuation on the basis of the "time use of facilities and labor" at every operation or step in the flow of materials through the enterprise.
- All product costs must flow into the inventory valuation system (raw material, work-in-process, supplies, finished goods) at all levels.
- All revenues are derived from the disposal of inventory.
- Overhead absorption variances derived from varying capacity utilization are profit variances and not product manufacturing cost variances.
- The cost system, manufacturing management system, and material management system should be tied together through a universal numerical manufacturing bill of materials and a matrix type part numbering and activity coding system.

7
The Information Structure in Manufacturing and Design

A. THE INTEGRATED INFORMATION SYSTEM

To develop a precision cost system for an enterprise and all of its activities, it is essential to first define the information and decision flow pattern, the required data base elements, their sources, and the areas of mutuality of management information and control. Figure 7.1 shows the relationships between the information used in engineering, manufacturing, marketing, finance, and materials management. This diagram demonstrates the need for smooth interdepartmental communications and a common information "language" for the interchange and multiple reuse of data and information by the various elements of the enterprise. This is especially important in the relationships between product design, manufacturing, sales, and marketing.

Many modern manufacturing firms try to apply "concurrent engineering." This is an organized effort to simultaneously design the product, its parts, the manufacturing process, and the required special tooling. The concurrent design team also often addresses the product's life cycle and its ultimate disposition and/or recycling. The product designer may use the "quality function deployment" (QFD) approach to address such issues as product performance, service life or survivability, quality standards, customer satisfaction, environmental impact, and after-use disposition. This process requires a team approach, the interlocking of key personnel, a common data base, good computer support, CAD/CAM software, and an effective data communications network. Such interlocking is not entirely new.

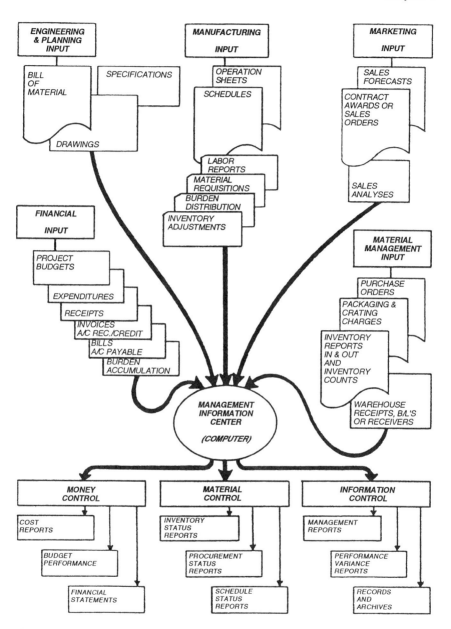

Figure 7.1 A generalized diagram of the management information flow in a manufacturing operation.

The Information Structure

Cooperative manufacturing engineering, product design, and sales effort is not a new idea. Designers and tool engineers have often worked together as teams in the past. Sales engineers have often consulted with designers to feed customer needs and desires into the design process. Manufacturing engineers and production executives have pressured designers to modify products for more economical processing.

In some respects, the information structure of a manufacturing enterprise is a "chicken and egg" situation. Successful sales and marketing activities require strong engineering and manufacturing support. The engineering function supplies ideas and product designs to marketing and manufacturing and also responds to their requirements. In most cases, the liaison is excellent and the products are both cost effective and customer responsive.

However, engineers sometimes produce designs which are difficult to sell and/or costly to manufacture. Sales people are sometimes rebuffed when they seek engineering support in their response to customer demands. Some managers are either unaware of these conflicts of interest or simply ignore them. The more effective manufacturing firms often build product teams and assign product managers to respond to these issues.

Design engineering develops the basic information and technical documents used to define and manufacture the product. These documents include design drawings, design and performance specifications, engineering bills of materials, and configuration control (engineering change) reports.

Manufacturing engineering supplies the manufacturing bills of materials, process designs, tool designs, cost estimates, manufacturing operations instructions (routings or operations sheets), and production facilities. Manufacturing also supplies the production cost and operational data used in making pre-design engineering estimates, applying value engineering techniques, and developing cost effective product designs.

These functions are a marriage! A manufacturing enterprise cannot divorce them and survive. The marketing and sales success of the firm depends upon them. The production scheduling and material control functions bind all these activities together and the cost accounting system tracks and evaluates their performance. The financial success (or failure) of the effort must be monitored and tracked. Precision cost measurement provides key information for wise marketing and operations management decisions, cost effective product designs, and economical manufacturing process development. Data base needs often interact, and success often depends upon the ability to rapidly access and retrieve accurate data.

Therefore, it is necessary to recognize the differences in the capabilities, and the similarities in the functions of manual and computerized information systems.

Although the objectives are the same, the capabilities of a manual system are very restricted. In a manual system, reusing information without rewriting is a difficult, slow, and error filled process, and data retrieval and analysis is cumbersome and slow. Photocopiers, blueprint machines, microfilm systems, multiple part forms, and ledger type accounting machines can help, but they are difficult to blend into a cohesive system and are limited in their effectiveness.

A manual product design system is also cumbersome and difficult to integrate into a "concurrent engineering" program. Documents need to be duplicated, marked, and modified. With photocopiers, photo drafting is feasible, but slow and cumbersome. Information must still be transferred by people, and the transfer is error prone.

A manual system can be used in a precision cost procedure. However, because of the required level of detail, manual systems often skip steps to expedite information flow, and leave gaps in the system to avoid cumbersome and expensive clerical procedures. This degrades the effectiveness of the cost system.

A computerized system with data base type software and graphics can rapidly and accurately capture, retrieve, rearrange, analyze, reuse, and report any item of data or any drawing or design. The system can deliver the data or drawing to the user or other interested personnel in any desired and appropriate format. The computer makes concurrent engineering more practical.

But, computers are quick and precise electronic "idiots." They can only follow instructions and can only respond or react to data according to their predefined programs or formats. The computer is not a "system." It is the tool which is used to make the "system" (information flow) more responsive. It can be used to integrate the system and coordinate activities like engineering, sales, and manufacturing on a "real time" basis for the benefit of the enterprise and the customer. But the result of this effort depends upon the system.

B. THE FLOW OF INFORMATION

The information system is composed of the codes, languages, documents, files, and decision criteria of the management process. In computer operations the system is defined in the form of programs and algorithms which support the information flow and decision process.

Among the most significant issues in the construction of a manufacturing and management information system are the design of the flow of information and the identity and documentation of data sources and users. There are five "basic truths" which must be recognized in planning a management system for computer application. These "truths" are shown in Table 7.1. As pointed out above, all product information is sourced from the engineering

The Information Structure

Table 7.1 The Five "Basic Truths" of Management Systems

1. The primary source for product information input into a manufacturing management system is the Engineering Department, and the vehicle for information input into the system is the manufacturing bill of materials.
2. Each time information is transmitted from people to computers, from computers to people, or from people to people, a potential error factor creeps into the system. Therefore, information should be put into the system in such a form that it can be used over and over again without risk of erroneous transcription or copying error. The system should be designed to minimize the error risk factors in data entry and calculations.
3. A management information system must have level-by-level and random data access and the information flow must run full circle. The system must feed back current operating and performance data on manufacturing operations, work-in-process inventory, sales, distribution, and finance in order to compare actual performance versus plans and schedules as the basis for management control decisions.
4. Estimates and prices are established before performance and the relationship between performance and profits is an historical function. With current information, more realistic and precise adjustments can be made in manufacturing and engineering programs to more closely approach the original estimate, improve profitability, and define the causes for failure to meet the estimate.
5. "Time is money" and this is manifested in terms of labor cost, inventory ownership cost, and the inventory volume required to accommodate the clerical lead time and the manufacturing schedules. As the system approaches true "real time" operations, inventory investment can be lowered and the precision of control can be improved. Thus, response speed is a major objective in modern management because response time is a competitive factor in all phases of management performance.

function in the form of drawings, bills of materials, and specifications. These source documents can be in the form of "hard copy" or computer files and images. The manufacturing bill of material (MBoM) is the key document for putting product information from engineering into the production management system.

In all business systems, the key records for the control of the enterprise are the accounts receivable and the inventory ledgers (Figure 7.2). Whether manual or computerized, all sales and financial transactions must flow into and out of the system through these records. All revenues flow in through the accounts receivable records and all customer based credit and delivery data should be stored in, or be accessible through the accounts receivable ledger. All sales orders should clear through accounts receivable for credit verification and acquisition of customer destination and shipping instructions.

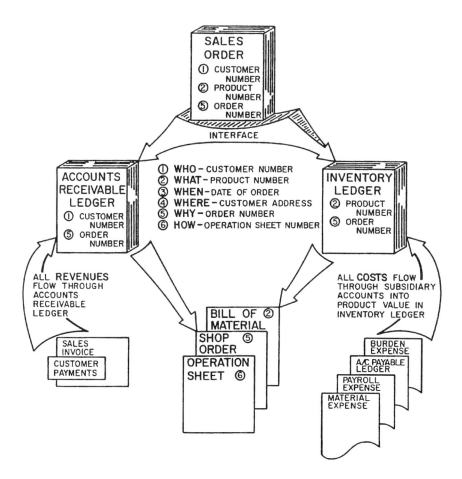

Figure 7.2 The data flow "story" and the relationship between the sales order and the accounts receivable and inventory records.

All receipts, work-in-process, and sales transactions must pass through the inventory records, and all expenditures and costs must be accrued into the value of the work-in-process or finished goods inventory. All product oriented management data should be stored in, or be accessible through the inventory records. Manufacturing and other operating costs can only be fully recovered through the liquidation and profitable sale of these inventories. Figure 7.2 shows how the sales order, accounts receivable record, and inventory ledger mesh into the development of the manufacturing information story.

The Information Structure

The information interface document between the marketing and financial activities and the internal product and manufacturing oriented activities is the sales order. It is usually the only document which contains both the product's identity and price along with the customer's identity and delivery information. It is not usually desirable to have customer information in the production system of a serialized manufacturing operation.

Customer identity at the shop level has a tendency to cause scheduling favoritism which limits the freedom to consolidate manufacturing processes and component inventories. The exceptions to this are in custom manufacturing operations where the contract or customer identity often follows the project through the design and manufacturing sequence.

An integrated management information system is based on the management concept of "Creation, Fabrication, and Control." The "Creation" element involves the creation of product, sales demand, and facilities. The "Fabrication" function converts materials into products and adds value. The "Control" function governs the top management, finance, personnel, production flow, materials control, and information control. These functions are linked together through the integrated manufacturing information system (Figure 7.3).

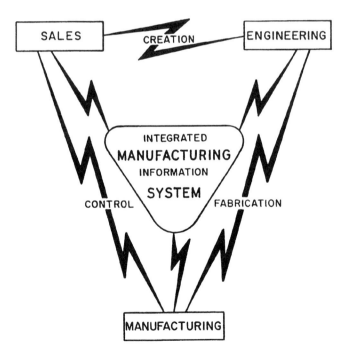

Figure 7.3 Interdepartmental information and communications patterns.

In general, the flow of cost information from the shop floor is based on a time and progress reporting system. Before the advent of computers, and in many currently non-computerized operations, the medium for shop order releases and progress reporting had been a job time card or "operations card." This document is being replaced by bar code systems and other direct wire methods for reporting to control computers in more advanced operations. This operations card, or its bar coded computer input replacement, is the means of reporting the beginning and end of any work task or material movement. While the replacement computer and bar code based system is widely used today, we will depict the data flow system using an "operations card" as shown in Figure 7.4.

In a bar code system, the worker wands his badge, the task menu, and the job order document number, and reports any scrap or rejects. The computer data file adds the other required information. In the operations card system, the job order data can be preprinted and/or key punched into the card, or entered onto the card by the worker. The time and date are entered by the time clock. The data content of the transaction is the same in either case, and these transactions gather the coded story as the work progresses through the process.

Figure 7.4 demonstrates the use of the journalistic check list as a part of the data collection for the operations "story." We need to know the Who, What, When, Where, Why, and How of every operation or transaction in order to report both progress and cost data into the system. Satisfying this requirement

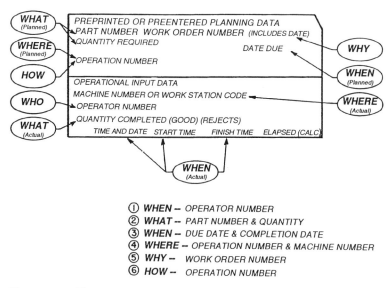

Figure 7.4 The operations card "story".

means coding all of the elements of the information system. We need operator numbers, machine and/or work station numbers, and operation numbers in addition to the part number, order quantity, and the time and date of the transaction. These numbers can be designed in the same format used in the SIMSCODER coding philosophy discussed in a later chapter.

Collecting direct and indirect manufacturing labor cost requires giving each employee an identifying number (often called a clock number). In some cases, the employee's Social Security number is used, but this has no interrelationship with the management information system and is very long. If possible, the employee number should provide useful, human legible information and, in modern systems, it should also be in bar code and magnetic stripe form on the employee's badge or identification card. The employee number ties attendance and wage rate to the operation or transaction in which the employee participates.

There are several technologies available to achieve this objective. Planners try to "hard wire" fully automated systems when possible. However, the state of the art in complex manufacturing operations usually requires some human intervention in the data collection activity. Automated identification systems reduce human error and work load factors, and can directly link the data collection station to the computer system. The most error free and widely used automated data collection systems are based on the application of bar codes.

The bar code presents a multidigit number in an optically (laser) legible format. Bar codes can be printed directly on a package, on labels to be pasted on the package or product, or on a tote box, product, part, document, or employee badge. The laser scanner can usually read the bar code in almost any position as it enters the item number and product identity into the computer memory for processing. At the same time, the computer can identify and enter the location of the scanner station and record the data entry time and date. This records the What, When, and Where of the transaction. If a worker is involved in the transaction, the Who can be recorded from the employee's identification badge. The location can be recorded, and a specific work station or storage facility identified.

The operation number and work station number cause the computer to report the Why and How of the data story from the work schedule and operations file. In other cases, the Why and How can be picked up by scanning either a transaction menu or a routing sheet with bar code legibility. The operation and scheduling data can also be on file in the computer and serve as an instruction source for feedback to a printer, LED display, or cathode-ray tube (CRT) at the work station. In automated and/or computer-integrated manufacturing (CIM) systems, the operations instructions to the machine can be retrieved from the computer file by the transaction.

In continuous process operations and/or computer managed Just-in-Time (JIT) systems, some of the data collection practices can be modified by the use of "travellers" or "KanBan" techniques. In these cases, variances in procedure or performance can be noted by recording material movement at transfer and storage points. In such systems, the labor costs are usually stable for each operation and the material utilization is controlled by the physical flow or materials handling system. The handling system can be equipped to report flow status directly to the computer. This reduces the human based reporting requirements.

There are also other, sometimes more sophisticated data acquisition and reporting systems. They all have similar functions, but use a variety of technologies in a wide range of applications. These include magnetic "touch" systems, optical and/or magnetic font readers, radio frequency (RF) systems, infrared (IR) reporting techniques, voice recording, and others. The bar code, however, is the most common and probably the most practical technique.

To explain the methods and benefits of the proposed system, we will suggest reporting and control system techniques which are capable of managing either a process industry, a batch production plant, a job shop type operation, or a serialized piece parts and assembly operation. Some of the reporting procedures can be eliminated by using handling automation and computerized flow control of the process. However, in order to develop a level-by-level, total absorption standard costing system with precision work-in-process inventory valuation, key control points must first be established.

The control points should be set at each work station where the product or part changes its basic identity or processing class. For example, the points of change are from material to semi-finished part, from semi-finished to finished part, from part to sub-assembly, and from sub-assembly to finished product. There are usually additional control points but these are the critical levels of inventory control. Changes in handling characteristics can also be used as an inventory or cost control point. For example, a change from loose handling to palletized units or vice versa would change the handling characteristics and the costs.

This text will use the following example procedure as the basis for discussion and description of the precision manufacturing cost system concept. This procedure is obviously subject to variations in order to suit an actual application in a specific manufacturing system. The data collection system would have the following features.

1. *Identification and data tagging*
    - The employees' photo-identification badges would be bar coded with their Social Security number, clock number, and permissible work areas. Badges would also have the Social Security number,

clock number, and permissible work areas recorded in a magnetic strip for use as a key card and/or telephone, computer, or photocopier access card.
- All purchase orders would have the purchase order number and item numbers printed in both human legible and bar code form. A computer terminal, printer, and bar code wand would be located at the receiving dock to produce receiving copies of the purchase order and to report items received. Bar coded purchase order number labels would be mailed to the vendor with the order and would be stuck on each package shipped.
- All operations or routing documents would have the applicable part or product number, work order number, and each operation number printed in both human legible and bar code form.
- All material requisitions would have their requisition number and material item numbers printed in both human legible and bar code form.
- All tote boxes, shop containers, and pallets would be identified by a serial number and bar code label.
- All store room items, materials, and finished goods items would be tagged or labeled with a bar code item number. Finished goods would also have a bar coded Universal Product Code (UPC) label.
- All tool room tools would have a tool number and be marked or tagged with a bar code label.
- All machinery, work stations, and storage locations would be bar code labeled with a machine, work station, and/or location number.

2. *Data capturing procedures*
   - Employee time and attendance reporting would be recorded by passing the photo identity badge over a multiple laser bar code scanner (as in a grocery checkout) or through a magnetic stripe reader on entering and leaving the premises. Employees in regulated or secured areas would use either a bar code reader or the magnetic key stripe to enter restricted work areas.
   - Employees' start and stop time on an assigned task would be recorded by wanding the bar coded photo identity badge and work order to record the Who, When, Why, and What for the operation.
   - Materials, parts, store room items, and documents which are transported in a "common denominator" tote box, shop box, or pallet would be wanded as loaded on the common denominator unit, and the container serial number would be wanded to tie them to it.

- Process progress would be recorded and labor cost assigned by having the employee wand his photo identification badge, the bar coded part or product, and the tote box, or pallet, to record the Who and What. He would also wand the operation sheet or router work order number for Why, and the operation number for How and Where.
- Material issue control would be accomplished by wanding the badge of the person picking up the material, the requisition number which releases the material from stores, and the work order number on the requisition or route sheet to charge the material to the job. Assuming computer produced requisitions, the quantity in the computer file would then be confirmed by keying in the issue count at the time of issue.
- Material purchasing control would be accomplished by wanding the bar code label on the incoming packages and the bar coded purchase order number on the receiving copy of the order. Where possible, the label would also have the item number bar coded and the order quantity marked on the package. This would be checked by counting, or accepting vendor count. Acceptance would be entered into the computer by wanding the item bar code on the purchase order receiving copy and/or the package.
- Burden or overhead charges would be entered into the system in one of two ways. By wanding the operation number on the operation sheet or routing, the assigned work station or machine would be entered into the system along with the worker identity, the material, the product item, the work order number, and the time. If the job is to be performed anywhere other than the planned work station, the operator would either key the work station number into the terminal or wand the identifying bar coded tag on the work station or machine. In either case, the work station number would cause the system to retrieve the "use rate" for the work station and thereby assign the corrected burden or overhead charges to the job.

In summary, all of the required data elements can be collected by automated or manual means to provide the necessary input to the data base. There remains the need to "tag" or code the sources and elements of information for ease of identity, handling, and manipulation

C. THE MANUFACTURING/NUMERICAL BILL OF MATERIALS

A properly designed numerical manufacturing bill of materials (MBoM) is the most appropriate, and often the most complete source for defining product

The Information Structure

configuration and manufacturing information input requirements. It can be designed to utilize an engineered systems language, to identify materials and components, and to provide a matrix for access to product design, product structure, component engineering, material control, and production control data at any logical point in the product manufacturing and materials control operation.

The numerical manufacturing bill of materials provides the matrix capability for total material requirements explosion, complete or modular part and component retrievability, and the quantitative analysis of product content and structure. It can be accessed on a downward vertical basis for product composition, content, bill of materials analysis, and materials requirements explosion.

On an upward vertical basis it can be used to produce "where used" files to identify the multiple product use of parts and to assist in development of design standardization. The "where used" listing will show material and component usage on a product line or enterprise scale. This helps to shrink inventory, reduce multiple designs, and assist in stocking analysis and production batching.

On a level-by-level horizontal basis, the MBoM provides the framework for level-by-level inventory management and production control systems. When applying a functional parts classification code in conjunction with a related geometric group technology code, the MBoM structure can be a key tool for identifying and retrieving similar and/or identical parts for standardization of designs, application of group technology based manufacturing procedures, and work-in-process materials management.

Since the MBoM is the information interface between product design and manufacturing information, it provides a structured basis for production management, materials management, and cost accounting procedures. By applying a unified part number and data code language, we can use the combination of the MBoM and the coding system to build a technical and management information retrieval matrix.

The high probability of recurring use of prior components in new product designs has made development of a functionally based matrix type part number coding and identification system very desirable. The matrix code system provides a means for focused retrieval of prior engineering and manufacturing experience. Additional product and manufacturing information is stored in either computer retrievable or document form, and its retrieval would be through the product and/or part number identification code.

Coding provides a human/machine legible language for use as a means for accessing archival data sources in product design, parts standardization, and product cost reduction programs. By utilizing a functional language code system, product and part design characteristics can be identified and captured as a by-product of the design process and computer compared with the design codes of the components of prior designs. Similar or identical design and operations information such as elemental design data, manufacturing cost statistics, design drawings, operations sheets, and complete or partial numerical bills of material can then be retrieved and reused to further product standardization, improve reliability, and reduce design and manufacturing cost. The same coding and data retrieval concepts can be applied to manufacturing data acquisition, sales analysis, market research, and financial controls.

Figure 7.5 shows a typical manufacturing bill of material structure for a complex manufactured product. This diagram depicts a manufacturing bill of material which is structured in manufacturing and assembly order. It should be noted that in the manufacturing bill of materials, the level of product completion is defined as an "indent" number. This is different from the engineering bill of material which is generally structured on the basis of engineering discipline or material class.

In all cases, this book, will be using the MBoM as the basis for discussion and consideration of costing procedures. To apply a matrix numerical coding system and a precision costing procedure, it is necessary to first establish a level-by-level manufacturing bill of materials.

Figure 7.5 and Figure 2.1 show that the process levels progressing from the raw material through final assembly can be coded with the same indent numbers as the corresponding bill of materials position or level of completion

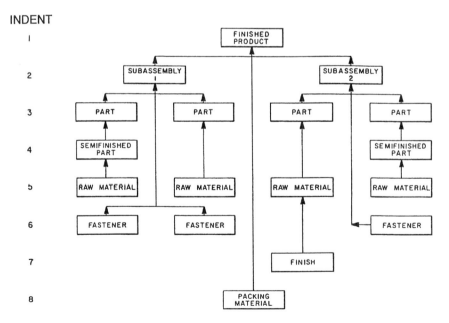

Figure 7.5 Typical manufacturing bill of materials indent structure.

The Information Structure

in the manufacturing bill of materials. Thus, in Figure 2.1, finished parts fabrication would be manufacturing sequence level four (4) and detail parts would carry indent level four (4) in the sample bill of materials. Likewise, the final assembly operation would be level one (1) in the operational sequence, and the indent code in the manufacturing bill of materials would show the final assembly or product as indent level one (1).

It is obvious that more complex products will require additional indent levels or steps in their assembly and manufacture, and simpler products will have fewer indent levels. Figure 7.5 has one less sub-assembly level than shown in Figure 2.1 and therefore its finished parts are at the indent three (3) level. However, the manufacturing bill of material's indent level codes should be company wide.

When this concept is carried to the next step, as shown in Figure 7.6, we can see that the generic or descriptive functional classification code, coupled with the indent level code, provides a specific identity for a class of information or items, and defines the level of completion as well as the item class. Because the indent code varies with the level of completion, the same item class number can also be applied to entirely different items in each level of the bill of materials. This provides a multiplying factor which, in these examples al-

INDENT		ITEM CLASS				
		501	502	503	504	505
1	FINISHED PRODUCT	STRAIGHT GEAR REDUCER	HELICAL GEAR REDUCER	HERRINGBONE GEAR REDUCER	BEVEL GEAR REDUCER	WORM GEAR REDUCER
2	SUBASSEMBLY	SHAFT AND GEAR	HOUSING AND COVER	BEARING AND CAP	BASE AND BOLTS	—
3	SUB-SUBASSEMBLY	GEAR AND SET SCREW	GEAR AND KEY	—	—	—
4	FINISHED PART	SPUR GEAR	HELICAL GEAR	HERRINGBONE GEAR	BEVEL GEAR	WORM GEAR
5	SEMIFINISHED PART	SPUR GEAR BLANK	SHELL GEAR BLANK	BEVEL GEAR BLANK	SHAFT BLANK	HOUSING CASTING
6	RAW MATERIAL	(CRS) STEEL ROD	(SS) STAINLESS STEEL ROD	(BR) BRASS ROD	(AL) ALUMINUM ROD	CRS STEEL PLATE

Figure 7.6 Indent levels versus item class codes as a data access matrix.

lows each item class number to be used six or eight times for six or eight distinctly different classes of material without fear of code duplication, confusion, or of exceeding the code's capacity.

A simple and somewhat "corny" method for defining the breakpoints in the development of indent levels is to: "ask the part what it is going to be next—if it doesn't know, then a breakpoint has been found."

For example, in Figure 7.7, if we were to ask gear blank 5-501, a semi-finished spur gear blank, what it was going to be next, its reply would be: "I don't know! I may be a herringbone gear, a helical gear, or a straight spur gear, but at this point I only know that I am a semi-finished spur gear blank."

We have, therefore, defined an indent level break point and an inventory level in both the level-by-level manufacturing bill of materials and the material control system.

For example, by using this numerically coded manufacturing bill of materials and parts and material classification coding system, we could inventory gear blanks to manufacture a variety of finished gears (or convert any other semi-finished parts to finished parts) from a lower cost and more homogeneous inventory level. This moves the work-in-process inventory investment upstream to a lower unit cost level with the resulting lower dollar lockup. This approach also increases scheduling flexibility by moving the final part item scheduling commitment downstream or later in the process. It also tends to simplify the application and implementation of both parts standardization and group technology objectives.

Since each part within an indent and functional classification group will have distinct elements of identification and function separating it from all other parts, it is necessary to break the indent-class group down with detail part identity capability. For this reason we will apply a distinctive and insignificant detail part number (often four digits or more) to each item or part which is different in any way from all others. This will provide a basis for a completely numerical manufacturing bill of materials or product parts list.

When this functional classification is combined with a simplified geometric shape classification—such as cylindrical, prismatic, or plate etc.—and further secondary geometry such as cross holes, hollow, or stepped—the part number code also contributes to application of group technology in the manufacturing process.

D. THE STRUCTURED INFORMATION MANAGEMENT SYSTEM FOR CODED OPERATING DATA AND ENGINEERING RETRIEVAL (SIMSCODER)

Many "package systems" have been offered for bill of materials processing, bill of materials explosion, inventory management, and cost accounting operations.

The Information Structure

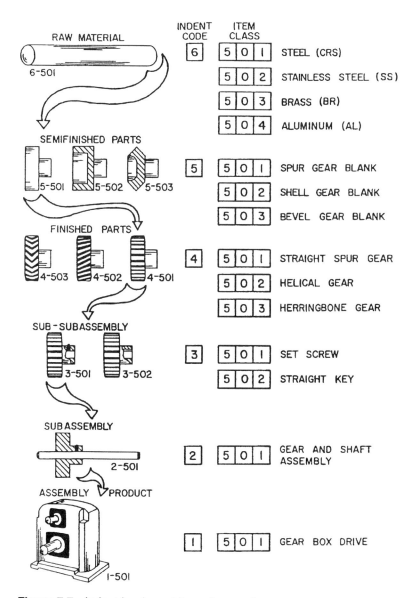

Figure 7.7 Indent levels and item class codes.

Most of these can function around almost any component, operation, document, or personnel identity coding system. In addition, many advanced computer spreadsheet programs can be used to manage materials, costs, and bills of materials.

However, to demonstrate the matrix concept of bill of materials management, detailed precision cost accounting, and manufacturing information management, we have chosen to use the "SIMSCODER" system as our medium of explanation.

Many other types of part numbering and information coding systems have also been developed. Some of these are all numeric and fully significant in their translation. Some are alpha numeric and others are semi-significant in their interpretation. Some also have a matrix information access capability. However, few of these coding systems envelop the drawing numbers, operation numbers, machine and work center numbers, employee clock numbers, and part numbers in a consolidated "code language" structure. Few of these systems combine their matrix codes with specifically designed numerical manufacturing bills of materials for random access to part and process information. A matrix coding system improves the effectiveness of information management and system integration. The "Structured Information Management System for Coded Operating Data and Engineering Retrieval" (SIMSCODER) component and information coding system has been specifically designed to support manufacturing and general management information requirements on an integrated basis. It can function very well in either a manual, punched card, or computer system environment. It is linguistic in its structure, but not language-dependent.

Figure 7.8 illustrates the relationships between contract or sales order numbers, component or part numbers and other coded management information. The significance of the part or component number is well demonstrated by its appearance on virtually every element of information input to the management system. Data can be accumulated on a part or component number basis, then manipulated by the information system to prepare specifically targeted output reports.

The core coding format (Figure 7.9) consists of three information access fields termed "indent level," "item class," and "item number." The indent level digit is used to designate the basic levels of a level-by-level material control system. The item class field, when coupled with the indent level digit, establishes a digital code which is referred to as a "component classification," or family group description.

The indent level information access field is the most important field in the coding system. The term indent level originates from the relationship of a particular component to its parent assembly and to the end product assembly configuration. This relationship may be effectively represented on a bill of material document by indenting successive levels of assembly according to product structure. The code levels of access are shown in Figure 7.10. The master manufacturing bill of material is often referred to as a numerical, or indented bill of material.

The Information Structure

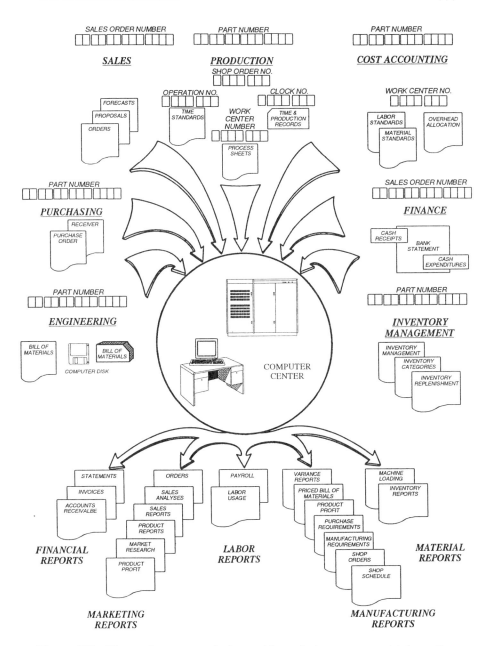

Figure 7.8 The system concept of a matrix coded management information system.

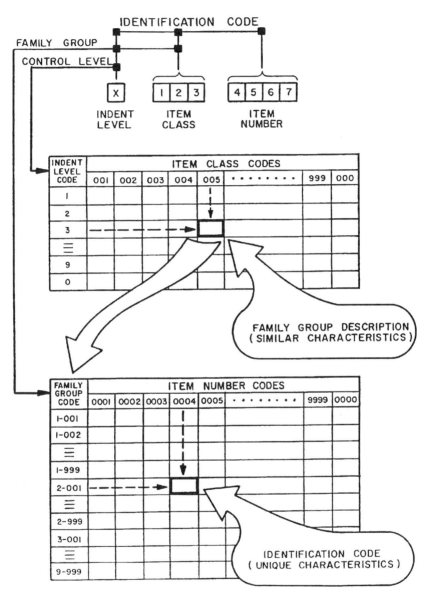

Figure 7.9 Matrix access levels.

The Information Structure

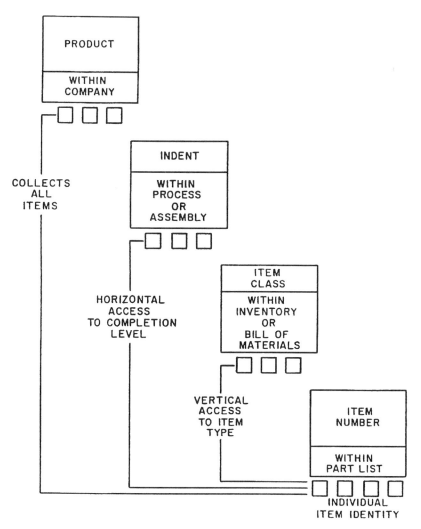

Figure 7.10 Code levels of access into the bill of materials.

This same technique can be programmed into a bill of material processor regardless of the type of item codes used. However, the proposed matrix coding technique simplifies the process through exploitation of its built-in logic.

In addition to identifying the material control level as a basis for effective materials management, the indent level code is similarly related to process control points supporting production control requirements. The bill of material structure relates to the process control points in the manufacture of finished goods (Figure 2.1).

One of the more subtle features of the indent level code technique is that, in addition to the usual inventory control points, there are other recommended in-process inventory control points between the entry of raw material into the system and the final exit of finished goods. From a materials management standpoint, the additional in-process inventory control points represent those stages in the process where the component or part changes condition or status. Early detection of unusual manufacturing performance is desirable and practical. The process can be controlled by management through a costed inventory record at each of these points, even if the material continues to flow through the system. The indent level code simplifies this procedure when properly applied.

The basic part number format is shown in Figure 7.11. This coding format can be expanded to include other levels of control and more items. The "basic line" field can identify a product line or a separate plant or division of a company. The "product group" could be a subdivision of the basic product line.

These same fields can be used for different data elements in labor and work station coding. They require the same matrix retrieval features. However it should be recognized that Figure 7.11 is only an example, and the number of digits in each field, and in the code as a whole, will vary with the application, the product, and the enterprise. The peripheral fields shown in the illustration can also vary with the application.

Although the core code, which is comprised of the indent level, item class, and random number, is all that is required to identify most parts in the bill of materials and on the shop floor, there is also a requirement for configuration control or engineering change management. Figure 7.11 shows a "variable" code position which can be used to denote the revision status of the basic part number. If the revised parts are interchangeable with their earlier versions, the part number remains the same with the added revision code. If the part is not interchangeable, it should be given a new part number. The revision code is a basic tool in configuration management operations. When a

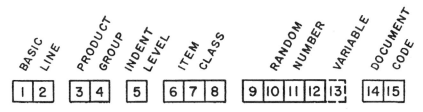

Figure 7.11 The basic SIMSCODER part numbering code format.

The Information Structure

part or product is progressively modified for improvement of the design or manufacturing operation, this code is used to define the sequence or status of the modified part or product. The initial design would use a zero in this position and subsequent revisions would be sequentially numbered. This also applies to sub-assemblies.

The "document code" field would be zeros in the case of a part, sub-assembly, or product identity code. This field is used to identify drawings, specifications, instructions, etc. which relate to the part or product with the same number. This field can also be used as a tooling code.

The same coding structure can also be used for identifying employees and work stations or departments. As an example, we can use the same fields for different data elements in labor and work station coding. This will provide the same matrix retrieval features.

If the part number code format is to be used for employee records and identification, the same coding fields could have different functions. In a multiplant or multidivision corporation, the "basic line" code in Figure 7.11 can be the identity of the plant or division where the employee works. In the same manner, the "product group" code can indicate the employee's department or "profit center" identity, and the "indent level" can be his work shift. These code fields would vary with changes in the employee's permanent work assignment.

The "item class" position can be a skill code identifying the employee's highest skill classification and pay grade. For example, a skilled tool maker would have a different code from a forklift operator or a janitor. Each would have a code for his highest employment skill in the company. Assuming the use of an in-place job evaluation system, the skill code would also identify the pay grade or classification of the employee. The shift code would also indicate the correct shift bonus or "for nothing" cost of the employee.

The "random number" would be the individual clock number of the employee, and the "variable" digit could indicate male or female, temporary or permanent, or some other personnel classification factor. The "document code" could be used to show the year of employment for seniority purposes. There are other possible configurations for coding the employee number and one alternative is discussed later.

All of these digits would be desirable in the employee record and in the bar code on the badge or identity card. However, only the skill classification and the clock number are required to report cost data from the shop floor. The other identity information would be available from the computerized data base in a bar code system and from the employee's attendance card in a manual system.

Therefore, as in the part number, this is a focused code structure. Only a portion of the code is required at the working level if sufficient backup is available in the data base or files. This employee code number answers the Who question in the data collection system. By comparing the skill code's wage rate to the planned skill level of the job, this code verifies the correct labor rate or defines the labor cost variance.

Another element in the labor cost data needs to report the correct operation, and the actual operation time versus the planned standard time. The operation's identity and time are also needed to compute the "use rate" cost of the work station.

The operation identity code must identify and classify the work station or function and the required labor skill classification for that operation. It should also provide the geography or location of the work station. Obviously in many cases, the geography of the work station will be complicated by multiple identical stations; therefore, a machine or work station number will also be required. The relationship between the actual work station number and the planned work station defines any scheduling variances based on the use of other than the specified operations' work stations.

For these purposes, the "basic line" code can identify the plant or division where the operation is normally performed and where its work station is located. The "product group" code can indicate the normally scheduled operation's department or "profit center" identity. The "indent level," used in the same manner as in the manufacturing bill of materials, can classify the level of the operation in the process. These codes are necessary for scheduling and cost analysis, but are not required on the shop floor.

The "item class" code identifies the type of machine or work station required and, thereby, the required skill level. The "random number" would be the sequential operation number. Positioning of the operation number in the code format is not critical if data base software is used. However, in a manual system, this field should be separated on the documents for ease of human recognition. These two fields are the primary shop floor operations code.

The "variable" digit would be used to identify engineering changes in the operation. The "document code" can identify the existence of instructions, drawings, or specifications concerning the operation.

To provide both the capability for capital equipment cost control and the ability to compare planned work station usage with actual usage, it is necessary to assign an identity number to each machine or work station. The same SIMSCODER format can be used again. In this case, the location, identity, and year of purchase are of interest. The code would be structured as follows.

The "basic line" would again be the plant or division identity, and the "product group" field would be the department or profit center where the equipment is located. In the case of machinery, these would be variable codes if the equipment is moved around the facility during its service life. These codes would depict only current location.

The Information Structure   117

The "indent level" would have the same meaning as in the part number and operation codes: it would represent the level of process completion where the equipment is assigned. This field could also be a variable code as the equipment is transferred from one use or location to another.

The "item class" would be used in the same manner as in the operations number. It would describe the generic characteristics or type of equipment as in lathe, milling machine, punch press, cupola, loom, etc. The same semantics issues must be addressed in assigning codes to manufacturing and office equipment as in naming parts and sub-assemblies. For example, one might have a lathe, threading; lathe, turret; lathe, mill/turn automatic; and lathe, vertical. Each would have the same primary code with a secondary variant.

In the machinery or capital equipment coding, the "random number" would be the purchase sequence number of the machine and the "variant" could show whether it was purchased new or used. The "document code" field whould show the year of purchase.

By using this code structure for capital equipment, one can monitor maintenance expense, service life, usage, location, and obsolescence for the machine.

The code would also be used in the manufacturing control system to relate and compare the planned use of the equipment in an operation with the actual use. It can become the key identifier for equipment cost and define the work station "use rate" charge. The machine or equipment number should be stencil painted on the machine and also show on a metal tag, along with the machine's serial number and vendor.

All of these code numbers serve as tools for the capture and control of information required in the manufacturing control and cost accounting system. These numbers tag the flow of material and information through the system and identify the sources of costs so that standard cost elements can be time related to the process. Figure 7.12 shows how these codes relate time and cost to a work station and aid in the collection of performance and cost data for input to the "operation story" (Figure 7.4), and for definition of variances from planned or standard performance.

E. LEVEL-BY-LEVEL MATERIAL AND OPERATIONS CONTROL

It has been stated that a level-by-level inventory control system is required as the basis for production control, and that without level-by-level inventory control it is impossible to have good production control. Let us examine this concept.

In a manufacturing or physical distribution system, the materials and work-in-process move from one work station or storage location to the next as a part of the flow of work through the enterprise. As each operation or storage is completed, the status, condition, ownership, or unit of control of the materials or work-in-process changes.

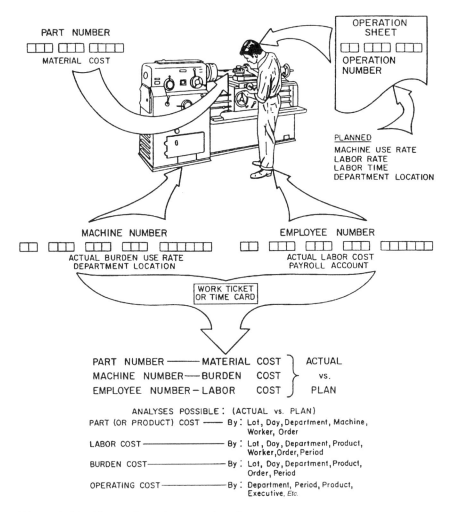

Figure 7.12 The work station cost data flow pattern.

For example, when raw material moves from a storeroom to the production floor, its ownership and location are changed and this is properly noted in the material management system. This is the first level or step in the process flow. When the material moves through a production process and is changed in its shape, condition, unit of control, appearance, or other distinguishing features, it becomes a different item from the original raw material. This change requires another notation in the material control system to assure proper recognition of the new status and identity of the material or the new part.

The Information Structure

This transaction generates a new part or identity code number. This is an inventory control transaction even if the material never stops moving. It must be removed from the raw material inventory record and entered into the work-in-process inventory in its new identity. As it moves from the semi-finished to the finished parts status it also changes its identity and part number. The identity code and indent level are defined during the process of building the manufacturing bill of materials, and the design of the product and its parts.

At each step, or level, in the manufacturing process, the product changes in some respect and takes on a new condition and identity. It may be converted from raw material to a semi-finished part, from a semi-finished part to a finished part, or from a part to a component of a sub-assembly or finished product. When assembled into a product or sub-assembly its individual identity is lost, and the new product or sub-assembly identity takes over in the material and cost control system. In each case the item's identity and condition is changed. This must be noted in the material control system in order to track progress, create an audit trail, and cumulatively collect cost information.

In a materials handling or physical distribution system, the same philosophy applies to changes in location and/or condition. For example, when an item leaves the packaging line and is loaded onto a pallet, its condition and handling characteristics are changed. This new transaction unit of handling must be noted in the material management record. The same is true when a carton is opened for order picking; the unit of transaction changes from a carton to an individual item unit.

Figure 7.13 shows the concept of level-by-level operations control in a simple manufacturing system. As the material moves through the process and changes its form, characteristics, condition, status, or identity, each change is represented by an indent level in the manufacturing bill of materials and a control or inventory point in the materials management system. While each level represents a control point in the materials management system, it is also an accumulation point for the collection of costs into the valuation of the work-in-process or finished goods inventory.

This level-by-level control concept is an essential element of the total absorption standard cost approach to precision costing of manufacturing operations. The costs are collected at each level to value the inventory in a cumulative process. The fully absorbed item cost is then available at each level based on the cost of manufacturing or movement up to that point in the process. This procedure provides both material control and cost control on a step-by-step basis for each item and each operation in the manufacturing system.

As stated above, the very existence of a level-by-level material control system is, by definition, a production control system. The manufacturing processing and shop loading procedures are scheduled to each material control level as production control points, and, if required, work-in-process banking points.

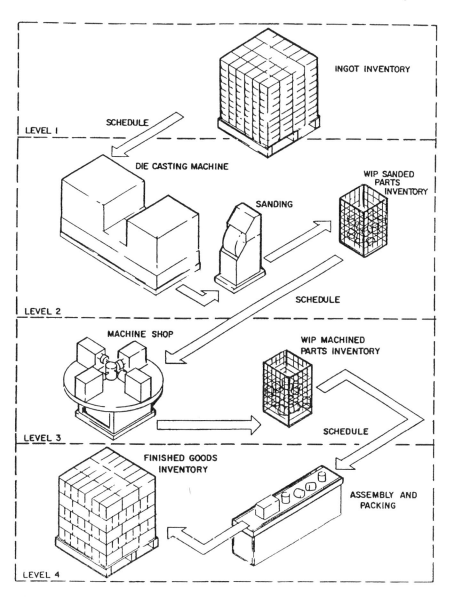

Figure 7.13 The level-by-level concept of manufacturing material and production control for a simple product.

F. THE MATRIX TYPE INFORMATION AND PART NUMBER CODING STRUCTURE

Effective communication within a manufacturing enterprise depends upon creation of a common "language" which is both human and machine legible. It must be capable of data handling and transmission among the organization's various activities.

To develop the coding system or language, the system designer must first establish a common human language and strict part and item definition semantics to assure uniformity and precision in translation. The system designer must avoid the creation of uncoordinated, and activity or department—dependent coding systems which might inhibit the mutual use of all available information by all elements of the enterprise.

If the advantages of an integrated data base and information processing system are to be realized, interfunctional language barriers must be broken and information must be homogenized and pooled. This can be accomplished through the application of an engineered numeric part number, bill of materials, and operations coding system with language characteristics, defined formats, and data base type computer software.

The coding structure must be functionally oriented to provide identification of information at all operations and management levels while maintaining data retrieval capabilities for use in engineering, material management, production control, cost control, marketing, and financial control.

The system language and its material and part number coding should tie the design drawing number and part number into the manufacturing bill of material. Developing a matrix type part number coding system is an essential first step in the design of computer based manufacturing management techniques and precision manufacturing costing and control systems.

The functional matrix type part numbers are basic elements in the development of manual, mechanized, or computerized drawing and document retrieval files and technical information retrieval systems. Good coding systems also simplify filing techniques, expedite engineering data retrieval, and help in both design and operations analysis. They also help the manufacturer manage product design and production operations. These coding techniques can also carry over into service parts support, reliability analysis, maintenance programs, and service parts manuals.

The cost system developed in this book is based on a matrix coding system developed by the author. This matrix coding system is similar in its format and function to the telephone number and the zip code (Figures 7.14 and 7.15). These codes have focusing structures which reduce the digit field as the code focuses onto the specific item's identity data. It leads the user from the

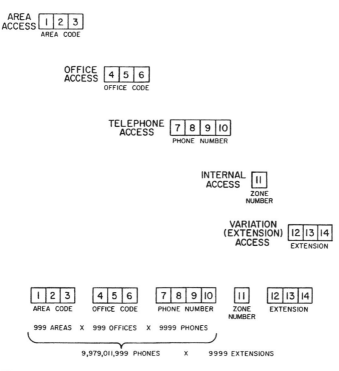

Figure 7.14 The matrix structure of the telephone number.

total or overall information label into the focused identity of an item or element of the information system.

In the telephone number, the area code is the first step in the focusing process. It is followed by the office code and then the individual handset or station number. At the office level, the area code is unnecessary. At the user level,

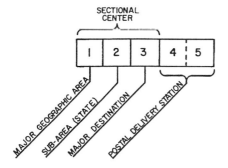

Figure 7.15 The matrix structure of the ZIP code.

The Information Structure

only the four digit line or station number is usually required. Thus, the structure focuses and simplifies its identification role as it reaches the information target.

The format or system described above is called a "Structured Information Management System for Coded Operating Data and Engineering Retrieval" (SIMSCODER). Its code sequence moves from the general data level in the manufacturing bill of materials to the specific part or information item. This same information structure can be applied in physical distribution costing as well as in manufacturing operations and product design.

It also has a built-in multiplier capability to provide maximum access and capacity with a minimum of digit positions, and is entirely numerical. This simplifies data entry and avoids problems caused by oral communication of alpha symbols with various linguistic accents and/or pronunciations. This use of numbers and avoidance of alphabetical symbols also allows for global communication without regard to the alphabets of other cultures' languages.

Other code structures can be used to label information and parts. Spread sheet techniques can also accomplish information stratification and sortation. However, because SIMSCODER was specifically designed for this purpose, it is particularly appropriate and adaptable for use in the proposed total absorption standard cost system and "use rate" method of level-by-level materials management and product costing. It is the result of many years of experimental development and successful practical applications.

SIMSCODER is custom engineered to provide access to product, manufacturing, and management information at all levels of product control. It combines a significant numeric matrix code, which identifies the functional elements of the product and operation at critical points of management control, with a random or non-significant detail part identity code. SIMSCODER provides a language-based, logical system for access to pertinent management or product information at any operating level or control point.

Figure 7.11 shows the basic part or product number structure usually applied in the SIMSCODER system. Figure 7.16 provides an example of the matrix access pattern which can be achieved with this code structure.

As stated above, the indent field represents the access or completion level and the level of the item in the manufacturing bill of materials. This field can have as many digit positions as needed to envelop all levels in the manufacturing bill of materials and the process. For example, if there are fifty-nine steps in the process and levels in the MBoM, a two digit field would suffice, but three digits would provide capacity for future growth to 999 steps or levels. Such complexity is possible in products like aircraft, ships, automobiles, etc.

The generic function code field is a polycode describing the generic function or name of the part, material, or sub-assembly at each level of its progress through manufacturing. At the raw material level, it describes the raw mate-

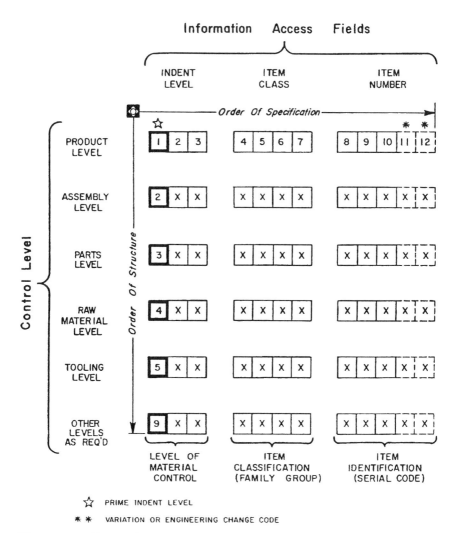

Figure 7.16 Typical SIMSCODER coding structure and format.

rial form. At the net rough shape level, it describes the final form (as in a casting) before any machining or operations are performed. At the semi-finished parts level it describes the blank part as in a gear blank or primary stamping. The finished parts are identified at the parts level by their generic functional names as in gear, shaft, bolt, handle, etc.

In each case, the semantics of the name is critical to assure the generic name's dominance over any secondary description or adjective which could change the meaning or identity of the code. For example, there is a significant difference between descriptions like "brick, fire"; "brick, paving" and "brick building." All of the "bricks" have the generic coding of "brick." However, the "brick building" should have the generic coding of "building, brick" versus "building, wood" or "building, steel." In all cases, the part or material title must fit the language and corporate culture of the user enterprise.

The Defense Logistics Agency of the U. S. Department of Defense and the General Services Administration of the United States Government have developed useful catalogues which address these semantics issues in a very practical manner.

An additional feature which can be included in the matrix coding system is the geometry of the part or material. This offers the opportunity to combine the item's function with its shape as the basis for applying group technology procedures. In such a code system, the shape factors which would be coded are the primary shape, secondary shape, internal shape, and auxiliary shape.

Typical primary shapes are prismatic, cylindrical, plate, and sheet. Secondary shapes could be none, partly cylindrical, stepped cylindrical, partly prismatic, trapezoidal, or formed plate. Internal shapes include such features as none or solid, hollow blind hole, hollow single diameter through hole, hollow multiple diameter, etc. Auxiliary shape would include none, cross holes, slots, threaded, flanges, etc.

These geometric features are coded in a separate field and applied to the part number as either a separate code block within the number or as a special "trailer" on the end of the code. Such a procedure allows the parts to be sorted by geometry into a rough group technology classification. The application of this group technology coding helps improve manufacturing costs. However, for the purposes of this book, we will merely cite the group technology possibility and leave its application to the discretion of the reader.

G. CONFIGURATION CONTROL AND DOCUMENTATION CODING

One of the most difficult materials management problems in a manufacturing system which supports technical or long service products is the matter of configuration control. This is the problem of dealing with engineering changes and their impact on inventory mix, field service support, and detail costing of the product.

When the new design part is completely interchangeable with its predecessor, the part number should remain the same; only the variation or engineering change code field should be altered. In such instances, the old inventory is usually worked out on a FIFO basis and the modified cost is addressed in the valuation of the inventory.

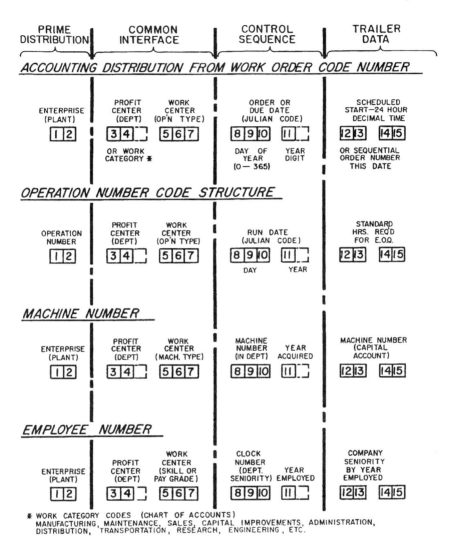

Figure 7.17 Examples of the application of SIMSCODER matrix coding to administrative functions.

The Information Structure

However, if the new part is not interchangeable with the old one and/or there is a recall and scrapping of the old parts for safety or functional reasons, the new part should be assigned a new part number, and the bills of material, inventory records, drawings, spare parts lists, and all other supporting documentation should reflect the change. In addition, the new part cost should be recognized in the product budget. Write-off of obsolete parts should be handled as a profit and loss item, not a product cost assignment.

In the best systems, drawings and technical data which support a product are coded to match or coordinate with the product's manufacturing bill of materials (MBoM) and part numbers. In such cases, the drawing number and supporting documents will have the same root number as the part or product, with a "trailer" code which identifies the type of document the number represents.

H. ADMINISTRATIVE AND OPERATIONAL CODING IN THE MATRIX SYSTEM

The same matrix coding philosophy can be applied to the design and development of administrative codes. This has the advantage of simplifying the programming of computers by providing built-in logic as a part of the code structure. It is also helpful if a manual or punched card system is in use. The codes are also more easily understood by employees. Figure 7.17 shows some examples of the application of the SIMSCODER matrix coding system to the administrative operation. These formats can be customized to fit a particular organizational requirement. However, the examples have been tested in practice and are usable without difficulty.

8
Material Cost Management and Control

A. MATERIAL STANDARDS FOR PROCUREMENT

Precise costing requires that materials standards become an element of cost control in manufacturing operations. The materials standards must include both quantity and quality. Purchased materials standards are also needed. Meaningful material procurement standards must include technical and performance specifications, physical dimensions and tolerances, and a definition of the required quality characteristics for the purchased materials, products, or components. The standards should also define the acceptable level of completion at the time of purchase, the standard or modular purchase quantity and/or delivery batch, the desired delivery schedule, the required packaging and marking, and the unit purchase price.

Many manufacturers who use "Total Quality Management" (TQM) concepts in their operations have extended their quality management policies and standards to include their suppliers.

There is also a growing trend toward the establishment of international standards. The European Economic Community (EEC) and the International Standards Organization (ISO) have established the ISO 9000 Standards. These are similar to and partially based on the standards of the American National Standards Institute's (ANSI) ANSI/ASQC Q90 series of standards and the United States Department of Defense standards (MIL-Q-9858A and MIL-I-45208). ISO 9001 deals with instrumentation, ISO 9002 specifies a model for dealing with quality assurance in production and installation, ISO 9003 specifies a model for quality assurance in final inspection and testing, and ISO 9004 provides guidelines on the key elements and the implementation of a quality system. In the future, all companies dealing with the European Economic Community will require certification of compliance with ISO 9000 and its components. American companies are also participating in this program.

The total quality management approach is aimed at eliminating quality problems at their source; this includes the vendors. TQM is a realistic management approach to achieving "zero defects" throughout the manufacturing system.

In many cases, the user/manufacturer can certify the suppliers' quality management procedures and vendors' quality standards. Suppliers can also be certified under the international standards. The suppliers' performance can be checked and monitored by making quality assurance visits to the vendor's plant and using sampling inspections of receipts. The proven reliability of the supplier's quality performance would preclude inbound 100% inspection, and in many cases, eliminate all receiving inspection.

Procurement standards should also specify the use of standard packs or quantities, and/or revolving reusable container systems. This assures the compatibility of the vendor's packages and batch quantities with the "common denominator" materials handling modules and work-in-process lot sizes of the materials management and costing system of the manufacturer.

In some cases, the manufacturer supplies the vendor with modular tote boxes, pallets, or special containers which fit the factory handling system's "common denominator" structure. The vendor packs these purchaser units in standard quantities and often attaches purchaser supplied automatic identification labels or tags with bar coded order numbers, part numbers, and quantities. In other cases, the vendor ships in customer specified packages which may be reused by the manufacturer for work-in-process handling or finished goods shipment. These practices tend to homogenize the materials handling system and support close material control and Just-in-Time operations from the vendor through the manufacturing system and finished goods shipping.

These practices not only improve the economics and control of the supply system, but also support ecological goals and policies by minimizing inbound packing material garbage. To save freight costs, the reusable containers are often designed for nesting or collapsing on their return shipment. They usually can make many round trips before their disposition or the recycling of their material.

These controlled vendor relations result in a stable system and a well defined procurement procedure. This gives the manufacturer a basis for development of good "landed cost" standards for inbound materials and components at the factory receiving dock.

In an effort to achieve these quality and scheduling goals at the supplier level, there is a growing trend in many industries for manufacturers to establish a working partnership with their suppliers. Purchasing agents are leaning toward the selection of material sources on the basis of certifiable quality standards, reliable scheduling, Just-in-Time delivery, and negotiated pricing. This approach requires close technical and managerial cooperation and often results in long term vendor/customer relationships. It also develops predictable standard material cost data.

In the past, most procurement was based on multiple competitive price bids from multiple sources. This often resulted in questionable reliability of quality, cost, and delivery, and little continuity of supplier relations.

The history of attempts to achieve quantitative management and control of inventories can be traced back to the turn of the century. After World War II, progress was made in the mathematical description of inventory systems. More sophisticated extensions of earlier theories were developed using operations research techniques. One of these developments was the "simple lot size formula," better known as the "Economic Order Quantity" or EOQ.

This EOQ formula was historically credited to Ford Harris of the Westinghouse Corporation in 1915. Later, several variations were developed by others including R. H. Wilson, whose name is associated with EOQ as the "Wilson Formula."

There are a variety of derivations of the Wilson Formula, and if it is rigorously applied to a specific enterprise, its logic is acceptable. However, the concept has several basic weaknesses. A major weakness in Wilson's EOQ formula is in the assumptions upon which the derivation is based. Most are not possible in a real life situation. The key assumptions are:

1. The usage rate is constant and independent of time.
2. No orders are split.
3. Items can be inventoried indefinitely and never become obsolete.
4. Costs remain constant.
5. Capacity limitations do not exist.
6. Ordering costs of one item do not affect the ordering costs of another item.
7. Storage space limitations are neglected.
8. Operating bottlenecks are neglected.

All of these parameters are variables. In real life situations, demand, procurement lead time, production rates, procurement costs, and ownership related costs are all variables. They all affect performance in different ways, at different times and in different combinations. Capacity, which is not covered by the formula, may be physically or financially limited by lack of plant equipment or available funding.

As the range and magnitude of these variances in the parametric values in the "real world" increases, the reliability of the EOQ formula decreases. This is a major reason why EOQ applications are only practical when applied to systems with relatively constant parameters.

When applied to production lot sizes, the EOQ approach becomes limited when items must be routed through different work centers. An economic lot size for one work center is often not economic for another. This condition may be overcome if economic lot sizes for one work center are even multiples of another, but this is not often the case. The EOQ concept also tends to contradict some of the advantages of flexible manufacturing systems (FMS), group technology (GT) and Just-in-Time management.

In a Just-in-Time management system, the lot size may be defined by the size of the material handling module or "common denominator" handling device, the inventory holding space at the work station, or the capacity of the materials handling machinery which supports the operation. In a flexible manufacturing system operation, the objective is often to achieve a lot size of one. That target would certainly not fit the Wilson test.

Determining accurate cost factors for use in the basic EOQ formula often presents another weakness in the approach. This weakness derives from the accounting system in use, not the formula itself. Accountants and industrial engineers seldom agree on the definition of "actual" processing costs. Although direct costs are often readily identifiable, indirect costs must also be accurately allocated to the process to develop true costs. An "engineered cost" approach is far superior to traditional accounting techniques for this purpose. The engineered cost or time use of facility method emphasizes the distribution and total absorption of indirect costs according to the actual manufacturing consumption of non-material resources. This technique is also capable of level-by-level process costing. If the EOQ formula is used, however, accurate cost data is essential.

The classic Wilson Formula, stated in words would be:

$$EOQ = \sqrt{\frac{2(\text{Annual Usage})(\text{Acquisition or Setup Cost})}{(\%\ \text{Carrying Charge})(\text{Unit Cost})}}$$

Obviously all of the terms of the equation are usually variables in a real life situation, whether in purchasing or in production. This raises additional cautions regarding its use.

Considering these problems, and the growing impact of JIT management methods along with often automated "common denominator" handling systems, the author believes the use of EOQ procurement or production lot size standards is an obsolete concept and should rarely be applied.

Instead, the standard unit of purchase should be a multiple of the standard handling module or "common denominator" quantity used in the manufacturing system. The vendor should be required to ship in modular transpor-

Material Cost Management and Control

tation units such as intermodal containers, truckload lots, or pallet loads (unit loads) whenever volume requirements justify such shipments. This type of delivery should be coordinated with JIT management of the operation, the design of the modular in-plant "common denominator" handling units, and the design of the inbound handling and storage systems. If the resulting quantities are rounded into multiple "common denominator" shipping and handling modules, EOQ may then become a valid measure for inventory or purchasing decisions.

B. QUALITY, SHRINKAGE, CHIPS, DROP OFF, AND SCRAP COST

Of the material cost factors that must be pinpointed for precision costing, the concept of quality is the most difficult to define. Modern management theory focuses on quality as a major competitive factor and cost reduction tool. Managers also differ in their emphasis on quality.

Authorities also differ in their definitions of quality. For the purpose of this text, we will define quality as:

> Conformity of the design and functional specifications, including adherence to composition, dimensional, finish, appearance, and performance criteria as defined in design specifications, and minimization of the deviation from these specifications.
>
> In the case of raw materials, this also includes adherence to alloying and formulation standards, weight and density criteria, and unit size, shape, and packaging specifications.

If the objective is Total Quality Management and zero defects, and the resulting profit improvement, the conformity of purchased materials and manufactured products to stated quality standards and specifications must be measured and monitored. Since 100% perfection is a desirable, but usually an unrealistic goal, the percentage of allowable error and deviation from standards must also be defined.

In the past, management set a target "Average Outgoing Quality Level" (AOQL) which was acceptable. In the zero defects mode, however, an AOQL objective of less than 100% good product is not acceptable. In real life, some errors must be anticipated and some scrap loss must be covered by inventory to assure reliable customer service. The TQM system must be based on definable quality standards, acceptable deviations from standard, realistic inventory policies, flexible capacity, an achievable service level, and practical monitoring criteria.

Some measure of shrinkage loss due to handling and counting errors, perishability, in-transit and in-storage damage and loss, and for some items, pilferage, must also be anticipated. None of these losses are 100% preventable, and they are probably not accurately predictable.

Some allowance must be made for shrinkage, even in the best run systems. Historical records and good judgement are probably the only means for making any kind of reasonable "guesstimate" of the extra inventory required to protect against shrinkage without degrading customer service.

For our purposes in this discussion, the standard direct material quantity will be based on the design standard material required for the planned method of manufacture. This will be M_D in the calculations. The planned design standard material is not usually the amount actually used. In most cases, excess material is required because there is some "chip" or drop-off of material in the process. Other material variances develop from the "nest and gain" effect and the need to provide for shrinkage and errors. Therefore, the computation of the actual material cost and the cost impact of errors and other variables is also necessary.

As stated earlier, using percentages for prediction of scrap loss is not valid because it inflates the work-in-process inventories and does not recognize the scrap impact of either the learning process or the relative difficulty of making the product. Therefore a method for calculating the predictable scrap losses which will flatten the growth of surplus scrap allowances is needed.

In the absence of good statistical data and probability analysis, the use of the square root of the scheduled production lot is an improvement. We can also apply suitable adjustment factors which recognize the favorable impact of learning and the negative effect of process difficulty.

Using the square root approach, we can develop a basic formula for calculating approximate primary scrap allowances in the absence of valid statistical analysis techniques. This formula might be:

$$M_S = \sqrt{n}$$

where: M_S = the quantity required for scrap safety stock
 n = the number of units in the production lot

To show the effect of this approach, let us compare its results with a percentage calculation for the manufacture of from 1 to 100,000 pieces. Let us use a 10% scrap factor versus square root.

| | Scrap | |
Order Lot Size	10 Percent	Square Root
1	1	1
10	1	3.1
100	10	10
1000	100	31.6
10000	1000	100
100000	10000	316.2

Material Cost Management and Control

With the exception of the small lots, this square root formula flattens the rate of scrap safety stock growth to a more realistic value.

The results from this formula must be modified by applying an "experience effect" factor based on the scrap learning curve, the required quality standards (AOQL) and the relative difficulty of the process. However, the learning curve is only valid where the work is labor intensive and the production runs are relatively long. When applying the learning curve for estimating manufacturing costs, the ideal circumstances should include the following:

1. Finalized part and product designs with completed specifications and bills of materials.
2. Long run production quantities and good historical experience records.
3. High direct labor cost content in relationship to material costs and machine time.
4. A good cost accounting system with labor distribution, material distribution, and production cost allocation procedures in force.

Machine paced and/or material or machinery dominated tasks do not usually respond to the normal learning curve predictions, and short runs do not have time to "learn." For example, CNC machines, punch presses, and screw machines "learn" to reduce scrap as soon as they are set up for production runs. Tool setting and programming losses are usually very small and predictable by machine type and product complexity.

If the scrap learning curve is used for material requirements predictions, the scrap allowance formula should be modified to respond to "experience." The process complexity can be recognized by applying a difficulty factor (D). This is a subjectively defined adjustment to the scrap allowance based on the combination of the ordinate to the scrap learning curve (f_S) and the user's experience or judgement. It is used to temper or adjust the scrap allowance to fit actual experience. It can also provide a tool for the adjustment of the scrap allowance based on process reliability and difficulty in machine based operations where the learning curve does not apply. It will also provide data for a continuous improvement program.

The difficulty factor is:

$$D = d(f_S)$$

where:

D = the difficulty factor in percentage.
d = the user judgement factor for the difficulty in percentage. (Trouble free items earn a 1% or 0% difficulty rating.)
f_S = the ordinate to the scrap based learning curve in scrap units per

1000 pieces at the current or predicted production volume. This is normally based on an 80% learning curve rate. In a machine based system where the learning curve does not apply, f_S would be one (1).

Thus, an approximate formula to calculate scrap allowances (M_S) in the absence of valid statistical analysis techniques would be:

$$M_S = d(f_s)/n \sqrt{n}$$

or

$$M_S = D/n \sqrt{n}$$

where:

M_S = the quantity required for scrap safety stock

The expected cost or quantity of scrap can usually be defined for a particular part or product. When that is possible, the predicted scrap cost or allowance should be included in the calculation of the cost or quantity of the required standard material. If scrap cost is not predictable, it should be treated as a variance and can find its way into the product cost through the "use rate" as a part of the manufacturing overhead.

C. "NEST AND GAIN" FACTOR—THE COST OF SCHEDULES

"Nest and Gain" costing is one of the most important, and most frequently overlooked material cost factors. The "nest and gain" factor (F_{ng}), is the material requirement effect of scheduling and lot quantity. This variance is usually generated by the different methods and sequences of production that are a function of lot size and schedule. The "nest and gain" effects are most noticeable in bar feed operations and in such "flat" part production as sheet, plate, cloth, leather, and paper pattern layout operations.

In the case of machining operations, as in a bar feed lathe or mill/turn center, the "nest and gain" factor is generated by the relationships between standard single piece stock length, standard bar length, and the width of the cut-off tool (Figure 8.1).

For example, in single piece production, a 3.750 inch standard piece of stock might be required for each 3.000 inch long piece part to allow for chucking and cut-off. Including stock for a 0.687 inch chucking end and an 0.062 inch wide cut-off tool, this generates a material "gain" of 0.750 inches per piece over the design length of the part.

If a 0.062 inch thick hacksaw cut-off is used to cut 69 pieces from a standard 22 foot bar, there will be an additional 4.216 inch hacksaw loss for

Material Cost Management and Control 137

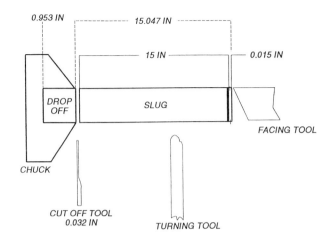

PRECUT "SLUGS"

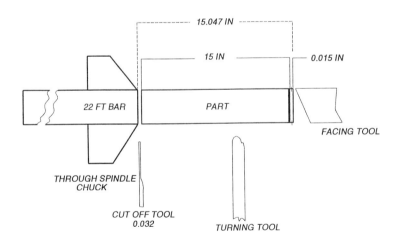

BAR FEED METHOD

Figure 8.1 "Nest and Gain" in a bar stock operation.

each 22 foot standard bar. This will result in a bar end drop-off of 1.034 inches for a fully distributed net material requirement of 3.765 inches per piece for the 69 individual 3.000 inch pieces made from the 22 foot bar.

Conversely, if the parts are made in quantity on a bar feed lathe using a standard 22 foot long bar, the bar could produce 86 finished parts based on a

0.062 inch cut-off tool with 0.668 inches left for chucking the final piece. This would result in a material requirement of 3.069 inches per piece, for a "nest" or stock saving effect of 0.696 inches less material per part than the standard stock per piece for pre-cut operations. This results in a 0.668 inch drop-off and chucking bar end and 86 parts from each 22 foot bar. The difference between these methods is the nest or stock saving.

Distribution of drop-off to the 86 finished pieces results in a net material requirement of approximately 3.070 inches per piece, for an 18.4% "nesting" factor. This is in contrast with 69 single pieces per 22 foot bar if the parts are pre-sawed into 3.750 inch blanks. The stock saving and nesting effect is 0.680 inches per piece for the bar feed operation, compared to single piece production. This saving is based on the choice of a lot schedule and lot size. It has no impact on the design or function of the part and only a small impact on the floor-to-floor machining time for the part.

The "nest and gain" effect is much more evident in the production of sheet metal or plate parts, the cutting of cloth or leather for clothing, and the layout of wood parts for furniture. In these operations, the arrangement of the parts' patterns in relation to each other in the cutout operation can have a heavy impact on the amount of surplus or drop-off materials. This is also a factor in the design of multiple part press dies and steel rule dies for paper and textile products. Figure 8.2 shows how the nesting of flat parts on a cutting or burn-out table can reduce the required amount of stock.

The ability to achieve this type of nesting is a function of both the shapes of the components and their scheduling in the shop. By combining parts into a single layout (which is only possible if they happen to be scheduled together and require balanced quantities), significant material savings can be realized.

This material variance is the "nest and gain" factor. Its omission from the planning process can result in an inflated raw material inventory and excess material cost. In stable manufacturing operations, the "nest and gain" factor is often predictable and it can be included in the materials planning calculations. When it is not predictable, most planners will use materials standards as the basis for requirements planning. In that case, the "nest and gain" factor will produce a variance in the material requirements and material costs. The impact of the unpredictable material variance should be treated as a cost of scheduling. It should be included in the factory overhead and be fed into product cost through the "use rate." In some cases, the material savings from nesting will more than offset the cost of owning the excess work-in-process inventory required to permit the nesting schedule.

When the scrap and "nest and gain" factors are predictable, it is feasible to compute the required adjusted standard direct material (M_{sd}) for each item or part on the basis of the following formula:

$$M_{sd} = F_{ng}[(M_D)n + M_S)]/n$$

Material Cost Management and Control

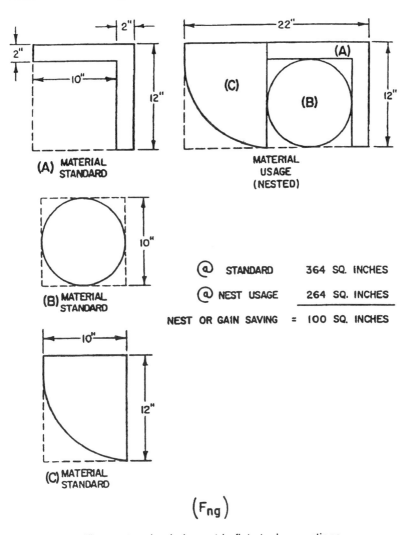

Figure 8.2 The nest and gain impact in flat stock operations.

where:

M_{sd} = the standard direct material for the part
F_{ng} = the nest and gain factor
M_D = the designed standard direct material
M_S = the calculated scrap allowance
$[d(f_s)/n\sqrt{n}\,]$
n = the scheduled production lot size

Indirect or contingent (M_{ic}) and burden (M_B) materials must also be accounted for in the calculation of the product cost. These materials are not usually assignable to an individual work station or product process. When they can be charged to a particular product, they might be treated as direct materials. In most cases, however, since they are non-specific in their use, they should be treated as part of the overhead or period expense and charged to the product through the "use rate."

M_{ic} = indirect or contingent materials

M_B = burden materials

D. INVENTORY AND MATERIALS MANAGEMENT

The above discussion confirms material management as the keystone element in all manufacturing management and cost control systems. Management's control of raw materials, work-in-process, and finished goods inventories is the control of funds flow through the manufacturing and distribution process. The level-by-level control of materials is the basis for production scheduling and a critical element of enterprise management. It is also the vehicle for cost accumulation in a precision cost system. Level-by-level material control is essential for implementation of Just-in-Time manufacturing and distribution. Level-by-level materials management and inventory control are fundamental components of the management philosophy presented in this book.

In the development of a precision manufacturing cost system, it is essential to recognize that:

All costs must flow into the valuation of the finished goods inventory and profitable disposition of inventory is the only means for recovering the manufacturing and distribution costs and the original investment in the enterprise (Figure 8.3).

A system for cumulative level-by-level inventory valuation is needed to achieve this objective. Two important general philosophies relate inventory and materials management to the manufacturing setting. These management concepts were emphasized in earlier chapters as the basic philosophy of level-by-level, precision costing of manufacturing and distribution operations.

They are:

- The level-by-level management of work-in-process inventory is the management of production flow.
- The control of material flow through the process is the control of funds flow and thereby the control of the enterprise.

The first steps in any materials management or manufacturing control program are the definition of the products and the development of a "master schedule"

Material Cost Management and Control

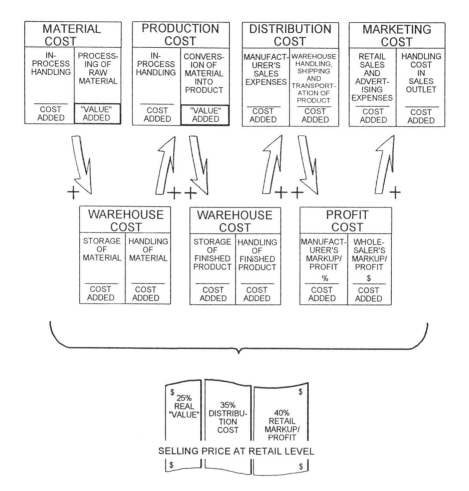

Figure 8.3 The flow of costs into inventory valuation and selling price.

so that all concerned personnel and organizational functions can focus on a rational enterprise goal. This requires documentation of the product design with suitable drawings, specifications, and bills of materials, and a projection of sales forecasts and manufacturing plans. The master manufacturing schedule is a planning tool which requires interpretation of sales forecasts and marketing programs into material requirements and production schedules. The master schedule is also the basis for development of the manufacturing budget.

Key factors in defining the master schedule are the production and materials "lead times" and scheduled "due dates." The parts' and components' due dates are derived from the master schedule and its product due dates. The manufacturing or purchasing start dates are "backed out" to accommodate the lead times for each part, component, and material item. In conventional operations, the flexibility of the lead time schedule is built into the WIP inventory projections. In Just-in-Time manufacturing systems, the WIP is minimized and the lead time projections must be more precise.

Obviously, the manufacturing process for each part and component must be defined and the operation times estimated to establish the required lead times for entry into the schedule. Vendor lead times must also be defined and introduced into the schedule. With these computed or estimated lead times defined, and the target dates for the master schedule established, the purchase or production start time for each item can then be calculated. Allowance must also be made for capacity utilization, shop loading, work-in-process ownership costs, dollar lockup, cash flow, and potential errors or variables in the performance of the process and the schedule.

In-house Just-in-Time scheduling of manufacturing can require excess capacity, or result in underutilized and partially idle production capacity. Full utilization of facility capacity and the use of "economical" production runs can also generate a work-in-process inventory buildup. Unreliable material and parts availability can also disrupt and/or stop production.

These are tradeoffs which have a cost impact, and all must be considered in the inventory management plan. All material, whether on-hand, purchased, or made in the plant, and whether scheduled for future or current use, must be accounted for and tracked in the system. All material must be valued by the cost system.

Different inventory items have different manufacturing process times or lead times, and different accumulated inventory values. Vendors also vary in their delivery schedules and reliability. This variability is a factor in management's make or buy decisions. It influences the decision to produce parts or components in-house Just-in-Time, to carry them in a work-in-process inventory, or to purchase them outside with a requirement for Just-in-Time delivery.

The control system must also have a mechanism for dealing with time and schedules. The system must manage the availability of items not yet on-hand and/or scheduled for later production or use. The procedure must incrementally accumulate costs and value the on-hand, work-in-process, and finished goods inventories on a level-by-level basis as the material progresses through the system. It must also track and account for the scheduled "on-order" materials not yet in the system and the "allocated" goods awaiting shipment or movement to the next operation.

To deal with the level-by-level system and the required lead times for the acquisition and/or manufacture of the materials and parts, it is necessary to use a time sensitive materials management system. A key factor in the design of such a system is the use of a four balance inventory structure with current and future time blocks or control periods. This structure establishes the level-by-level movement of inventory and the times or due dates of the transactions. Each time block or control point should have four inventory balances. These four balances are:

Balance on-order (0 to +) or due in from prior operations or vendors.

Balance on-hand (0 to +) or actual material in the plant warehouse, stockroom, or work-in-process. This is the dollar lockup. This is the inventory balance which collects the costs from the precision cost system. Its valuation is shown on the balance sheet.

Balance allocated (0 to −) or committed to be used in a subsequent operation or sale to a customer. This balance may include material on-hand or on-order.

Balance available (− to +) or the algebraic sum of the on-order, on-hand, and allocated balances. This inventory balance is the basis for scheduling new purchases and manufacturing operations. It is the controlling balance.

These four balances must be recorded in the inventory record at every control point in the materials management system. As each item is received, or moves from inventory to an operation, from operation to operation, from production to the warehouse, or from the warehouse to the customer, the transaction must be recorded (posted) to the proper record and all four balances must be suitably adjusted. Figure 8.4 shows an example of a purchased part inventory record using the four balance format.

In cases where the item changes identity, as in the change from raw material to semi-finished part, the source inventory must be vacated and the value and quantity entered into the next inventory level. In most cases, this requires the assignment of a new identity number or code.

Only the on-hand inventory is actually valued as an asset.

The valuation is developed through the accumulation of costs at each prior step in the work flow. The on-hand inventory must be valued after each transaction. The on-hand inventory values can be applied to the on-order and allocated balances to project cash flows.

In most systems, and in the proposed system, the inventory valuation will be based on standard costs with a comparative actual cost for manufacturing management purposes.

These transactions must be projected out into the future on a period basis to anticipate and accommodate the lead times for scheduled work and incoming materials. These records are "snapshots" of the status of the operation at each transition or control point in the material flow through the enterprise and in the predicted future performance pattern.

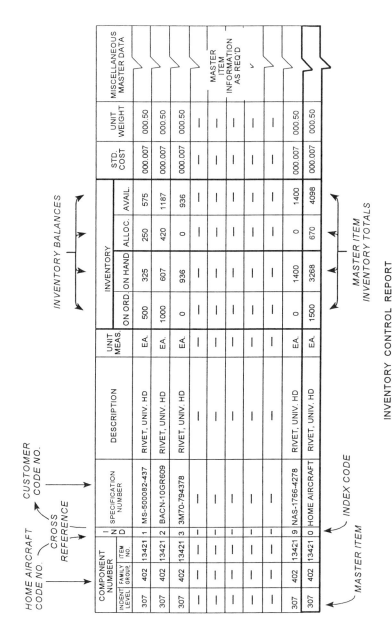

Figure 8.4 A typical four balance inventory record.

Material Cost Management and Control

The control system should be designed to adjust and update data continuously. The adjustments can be hourly in a computer based, JIT, real time system. More realistically, the adjustments are daily or weekly. In the past, with manual systems, they were often monthly and therefore seriously lagged the actual performance. In most cases, the adjustments produce a rolling or cascading adjustment to each current and future balance.

In some cases, with long lead time products, these systems project material requirements months and even years into the future. This procedure assures the inclusion of time in the material management process. Even in a long lead time situation, the precise control of inventory is desirable and in most cases, it is required to achieve Just-in-Time production.

Although the on-hand balance is the only one included in the financial statement, the other balances can be costed on a standard cost basis for planning, forecasting, and cash flow analysis. For example, a costed on-order balance defines the purchase and production commitment and future capital requirements on a scheduled cash flow basis. The costing of the allocated balance defines the value of backlogged production schedules or customer orders.

Such systems are not new. Periodical and predictive material and production control systems have been operated manually for decades. With the advent of computers and the development of software programs and "packages," most material management systems have become known as "Materials Requirements Planning" or MRP systems. They are now an integral part of most manufacturing management systems. Standard software packages are available for the application of these methods and procedures on the computer. Some of these software packages include sophisticated mathematical algorithms for materials management. These tools improve the precision of control and forecasting. The applicable computer packages can expedite accumulation of the "real" costs of manufacturing and distribution into the valuation of the WIP inventory.

The ability to project the master schedule into the future and to convert it into a detailed production plan, is the usual function of a MRP program. This projection capability is the basis for tight planning and control of WIP inventory. It is the foundation for production management and it is essential to the application of JIT techniques.

Materials/inventory management is the key element of a cost analysis and control system. Coordinated procedures must be established to support the interrelationships among materials management, production management, and the precision cost system.

This begins with the definition of materials requirements and costs based on the manufacturing bill of materials (MBoM) discussed in Chapter 6, Section C. This document defines the quantity of raw material and the number of units of each manufactured or purchased part required to produce the product. When using a pre-costed bill of materials or a product budget, the standard cost of each material item will be known and the total standard product cost can be computed. When planning for the manufacture of a new product, standard costs and engineered estimates are required to establish the part costs and develop the product cost budget. Figure 8.5 shows some of the information required for the compilation of a pre-costed manufacturing bill of materials.

The manufacturing bill of materials must be all inclusive. It must recognize and record all major parts and components. It must also list common items such as fasteners, finishes, labels, lubricants, and bonding agents to assure their inclusion in the cost of the product. All items in the MBoM must also have a part number identifier which is based on the matrix coding system.

The next step in the materials requirement planning process is the explosion of the manufacturing bill of materials to compute the numbers of each inventory item, part, or component required for the scheduled production lot.

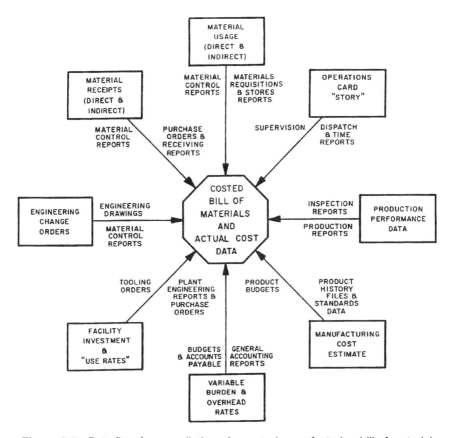

Figure 8.5 Data flow for compilation of a costed manufacturing bill of materials.

Material Cost Management and Control

This process, discussed in Chapter 6, Section C, requires multiplication of the quantity of each part or material unit for one finished product unit by the number of product units scheduled for production.

If the parts have been made or purchased before, the lead times will be defined and entered into the system. If it is a new product or part, the manufacturing engineer will define the process and estimate the required lead times.

The master schedule will be used to generate an overall production schedule for each product and establish the lot sizes or batches of finished product required to respond to customer expectations. This procedure will also relate schedule requirements to capacity and adjust the schedule or the capacity required to meet customer needs. It is at this stage in the planning that a decision must be made concerning the alternatives of Just-in-Time scheduling or full equipment utilization and work-in-process inventory.

Historically, most manufacturers had consolidated materials, parts, and components requirements and scheduled "economical lots" for purchase or production. This usually resulted in more fully utilized facilities and a build up of work-in-process inventories. The cost impact was an apparent reduction in distributed overhead or burden expense to the product and a higher dollar lockup in work-in-process and materials inventory.

Since the introduction of automation, flexible manufacturing systems (FMS), and the Just-in-Time philosophy of manufacturing, the objective has been to reduce lot sizes and work-in-process inventories. This trend leads to the elimination of the "economic lot" concept of scheduling and reduces work-in-process inventory. It also tends to reduce the utilization of manufacturing equipment by replacing work-in-process inventory with excess production capacity and shorter response times.

Following definition of the required quantity, and scheduling the due date for each production lot, lead times must be applied to back off the required due date to the starting date of the manufacturing process for each part or component. These dated requirements must be entered into the inventory record in their scheduled time blocks as an on-order demand or allocation. The machine loading or shop floor scheduling can then be started.

Some work-in-process inventory and some idle production capacity will be generated to achieve an optimum balance between capacity utilization and inventory dollar lockup with as nearly a Just-in-Time material flow schedule as possible.

In the design of a manufacturing schedule, and in the development of a manufacturing cost budget, it is unwise to assume full utilization of the facility and its production capacity. To assure enough flexibility to absorb unscheduled down time and other delays, a practical shop load factor of 80% of capacity can be used for the initial loading. This factor should be periodically revised to a high ninety percent (90+%) loading for the current week's work schedule. This load factor should allow enough capacity flexibility to handle scheduling changes, order cancellations or additions, absenteeism, maintenance down time, and other unplanned variables without failure to support customer service objectives. The standard cost should include an allowance for this capacity "cushion." Achievement of Just-in-Time operations in the face of these variables requires a zero defects program, a high level of preventive maintenance, some excess production capacity, flexible worker skills, and some work-in-process inventory.

As discussed earlier, each manufacturing operation or movement of material must be recorded in the inventory "level" record. The recording or posting of the inventory record at each transaction point can be accomplished using a variety of means.

The initial scheduling and inventory data is entered through the computer keyboard or as a by-product of the overall scheduling operation. In the continuing manufacturing operation, the most practical of the modern techniques is the use of a bar code wanding system to enter the transaction, operation, location, operator, date, and time. A keypad or telephone keys can be used to enter the quantity. In some cases a wire connected counting scale can be used to record the quantity. It is also possible to wire connect machinery to the computer for automatic recording.

As each production operation or movement transaction is recorded, the materials requirement planning software program should cascade the updating of all related transactions at every level throughout the system. At the same time, the accumulated costs are applied to the work-in-process inventory. It is important to maintain system discipline and to record every transaction in order to assure currency and accuracy.

E. WORK-IN-PROCESS INVENTORY AND JUST-IN-TIME OPERATIONS

Work-in-process inventory is either the material which is actually being worked on, or the surge inventory between varying production rates in sequential operations. Surge inventory can also result from batch production schedules aimed at the achievement of high levels of machine loading. The upstream inventory held by a supplier is also a work-in-process inventory.

Work-in-process inventory can also sometimes be used to reduce the total inventory investment. By holding lower value, semi-finished parts and products which can later be finished in a variety of ways in subsequent operations, the manufacturer can reduce total dollar lockup, reduce sales order lead time, improve customer service response, and improve the flexibility of the work schedule.

This practice is also applicable in situations where upstream operations require hard automation and expensive setups while downstream operations are more responsive. For example, in the manufacture of decorated glass tumblers, the unpacked blank ware can be stored to await customer dictated decoration and packing on a quick response basis. It is also a factor in dealing with partially processed food products which have a variety of finished recipes. For example, tomatoes can be processed into puree in season and stored for later packaging as ketchup, spaghetti sauce, or soup.

In all cases, the cost of work-in-process ownership, handling, and storage must be treated as a process cost and accumulated into the final valuation of the finished goods inventory. It is a part of the product cost.

In the past, the thrust of management thinking has been to achieve maximum utilization of machinery and to accept the buildup and ownership cost of work-in-process as a necessary result. The WIP was also accepted as a scheduling cushion which made the machine loading process more flexible and the assembly support more reliable.

Work-in-process inventory serves as a surge tank to break the lockstep between operations and allows more flexible scheduling of production. In contrast, Just-in-Time operations require close coordination and balance between operations with a minimum of WIP inventory and sometimes an excess of upstream production capacity.

The concept of minimum work-in-process inventory or Just-in-Time material management is not new. It has been partially practiced in the bottling and automotive industries for many years with some success. Bottlers have insisted on regular and scheduled vendor delivery of bottles, cans, and cartons and have often reduced supply inventories to a few hours. In such cases, the supplier has sometimes been penalized for shortages and forced to keep in-plant inventory and on-site vehicle loads to avoid packing line shut down. The automotive industry has tried to use the same techniques with suppliers and upstream plants. In most cases, these practices have forced the work-in-process inventories upstream to the supply point and increased the amount of in-transit inventory to assure reliable delivery in spite of weather or traffic problems.

The Just-in-Time concept became more popular in the 1980s because of its successful application in Japan. The logic of "streaming" the movement of materials through the system and reducing the bulges of work-in-process became obvious to even the most skeptical of western executives. The problems of its application were not so obvious. JIT's limited documentation also complicates the capture of actual cost data for precision costing. At the same time, the modularity of JIT material flow control simplifies the application of engineered standard costs.

In a JIT system, the balancing of production volume between operations becomes a critical factor. This often tends to increase idle machine time in order to provide sufficient machine capacity and flexibility to match the fabrication rates with the JIT usage rate. This can cause the addition of extra upstream capacity and idle time to reduce WIP inventory. It also encourages the use of outside vendors and their upstream inventories to support Just-in-Time operations. Each of these WIP inventories and/or idle capacity units generates a cost which must be included in the valuation of the work-in-process and product.

As stated earlier, it is the author's belief that the module of material movement and work-in-process storage within a Just-in-Time operation should be one or a multiple of the plant's material handling system's "common denominator" handling modules. This modular approach helps to homogenize the handling system and makes the physical characteristics of the individual parts and product components "invisible" to the handling system. It also simplifies material control procedures, provides a standard for handling and storage costing, and assures the flexibility of the system in the handling of a variable mix of parts and products.

The backup stock at each work station should be two or more modular handling units of the input parts or materials. This practice assures the continuation of the operation while the empty module or container is being replaced by the supply system.

In a well designed JIT system, the work station supply operation can be automated and paperless. If a manual system is used, the KanBan approach to "pull" replacement is an effective tool which avoids the use of paper but uses a "document" to trigger the replenishment cycle.

In an automated system, the input holding fixture or the conveyor feeding the work station can be equipped with a signal mechanism which can call for replenishment. This can be a bar code reader, a scale, a counter, or a signal from the transfer or loading device on the work station. In such cases, the signal would call for an automated guided vehicle, a conveyor, a turn table, or a robot to take away the empty "common denominator" container and replace it with a full unit. Such a system should also report the transaction to the material control computer and the cost system. In some cases, the signal might activate a signal light or be transmitted to a radio dispatcher, and the transfer would be conducted by manned vehicles.

The whole objective of the Just-in-Time concept is to reduce the dollar lockup in work-in-process and to free up the capital for other uses. Financial fluidity is a primary goal of Just-in-Time operations. When the operation is properly balanced, Just-in-Time techniques improve work flow and facilities and labor utilization.

F. INVENTORY COSTING AND VALUATION

As stated above, all manufacturing process expense, overhead, and inventory ownership costs must be absorbed into the valuation of the inventory on a progressive and level-by-level basis in order to accumulate and define the true cost of the finished product. The total absorption standard cost approach presented in this book uses the time use of facilities "use rate" method to capture the indirect expenses and adds labor and material cost to generate a precise and volume independent level-by-level inventory item unit cost. A generalized model of the procedure is shown in Figure 8.6.

This level-by-level inventory valuation approach has the advantage of defining a fully absorbed cost for each part or work-in-process item at every stage of the flow through the manufacturing process. This process also helps to provide some financial control parallel to the inventory and production control. It also provides information on the allocation of funds committed to work-in-process inventory.

The time use of facility or "rental" approach which is offered in this text properly allocates the cost of overhead to the product on a time use of facilities "use rate" basis. Idle time and overutilization of production equipment are treated as variances and as adjustments to profit. These variances are not direct product costs. They are the cost of scheduling. They can be reallocated to the product cost as part of the overhead and are included in the "use rate" charge.

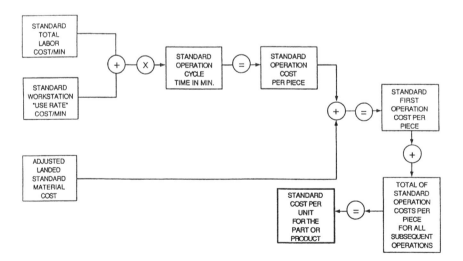

Figure 8.6 A generalized model of the development of a part or product cost.

The computed overhead cost of manufacturing each item is based on the standard "use rate" charge for the operation. The "use rate" is automatically self adjusted to account for any improvements, changes, or failures in the production process which result in a different operation time or a change in the assigned work station. An operation time change will modify the total "use rate" charge. A work station change will alter the basic "use rate" because of different floor space and equipment values. In the case of an unusual or occasional work station change, the difference in "use rate" charges (and probably labor costs) will be treated as a production cost variance in the profit and loss statement. It will be fed back into the the manufacturing cost as a periodic adjustment to the "use rates."

The cost of idle capacity should be charged to the marketing operation which failed to sell the capacity. When sales surges result in short term overabsorption of overhead in excess of the standard, the variance should be credited to the improved sales performance.

G. GETTING THE MATERIAL COST INTO THE PRODUCT BUDGET

The valuations of raw and purchased materials inventories are based on the landed cost of these materials at the receiving dock plus any allocated storage and handling charges which may be accumulated in their movement to the first production operation. At that point, the total material cost is entered into the work-in-process inventory with all of these costs fully absorbed. Materials are only charged into the WIP at the first operation where they are used. However, the storage and handling cost of WIP is part of the accumulated product cost. If it can be defined in relation to a specific operation or product, it can be allocated to inventory as an accumlulated operation cost. If it is mixed with other products, it will find its way into the product value as part of the factory burden portion of the work station's "use rate."

From the first operation, the costs of direct labor and the applicable "use rate" charges for the first operations are added to the total material cost to compute the fully absorbed valuation of the work-in-process item following completion of the first operation. This value then becomes the base cost for the next operation. The direct labor and "use rate" charges for each subsequent operation are added incrementally to produce the fully absorbed WIP cost at each stage.

Little attention has been given to the cost of storing and owning the WIP inventories or the financial impact of the dollar lockup that these stores represent. In many cases, a minor improvement in work-in-process turnover can generate a massive improvement in working capital. This emphasizes the con-

Material Cost Management and Control

cept that *money is the only material of production which earns money when it is in idle storage* (in the bank or in investments).

Inventory always costs money to own and must be sold to become fluid or spendable. Work-in-process inventory is a pile of unspendable dollar bills! In most cases, WIP costs money to own. This cost must be added into the cost of the product.

The value of each item in work-in-process inventory, or in finished goods, is therefore the accumulation of the material, labor, and "use rate" charges for the first operation plus the incremental labor and "use rate" charges for each subsequent operation up to the stage under consideration or throughout the manufacturing process. The WIP storage and handling costs should also be treated as an operation for each WIP "bank" in the production sequence. Expressed in mathematical format (with the elements values of materials, labor, and "use rate" omitted from the formula) this would be:

$$I_C = (M_{sd} + S_{hc}) + [(L_D + U_R)_1 + (L_D + U_R)_2 + (L_D + U_R)_3] + \text{etc.}$$

where:

I_C = fully absorbed inventory unit cost or value at any stage.
M_{sd} = adjusted standard material including scrap and "nest and gain" allowances.
S_{hc} = storage and handling cost of WIP and finished goods.
L_D = direct labor cost for the operation.
U_R = "Use rate" cost for the operation.

In each case, the unit cost of each element must be entered into the formula to value the inventory. The basic material cost would, of course, be derived by adding the purchase invoice plus inbound freight charges. The methods for computing the value of the labor and "use rate" will be discussed below.

9
Labor Cost

A. ORGANIZATIONAL PAYROLL PHILOSOPHY

Labor costing depends upon the structure of the workforce, which is a direct reflection of the organization's human resource philosophy. With very few exceptions, manufacturing companies use different compensation plans for management and staff as compared to factory or production labor. In most cases the management and staff are considered "salary" personnel and receive different pay rates and fringe benefits from the "hourly" or production people. When the production people are in a union, the salary people are usually excluded from the union contract and often have a different benefit package from the union members. In some cases, the supporting clerical staff is in the union, but supervisors and managers are usually not.

There are some exceptions, however. It is not unusual in the aircraft and other large high tech industries to find that the engineering and technical staff have their own union. Also, teachers and university professors, hospital technicians and nurses, and public employees are often members of specialized unions or special locals of the mainstream unions. Supervisory personnel at the shop floor level are sometimes included in the union contract in heavy industries such as steel, chemicals, and mining.

There are three basic methods of paying manufacturing labor personnel, and many individual company variations of each of these methods. The most common pay structures are the hourly wage, the weekly or monthly salary, and some form of incentive wage plan.

Most employees also receive fringe benefits such as paid holidays, vacations, and health insurance. The employer's payroll cost also includes the employer's share of the social security tax, workers' compensation taxes, and unemployment compensation taxes.

In most industries, straight salary is used for management and staff personnel with some form of annual or periodic bonus plan or incentive for key management executives. Salaried staff personnel also sometimes receive bonuses or incentives, and most clerical and supporting staff personnel are paid for overtime.

In most industrial enterprises, the hourly employees are paid a "call-in pay" of four hours' wages if they are called in and then do not work the scheduled hours. This call-in pay should be treated as a "for nothing" cost and be included in the overhead calculation. The employees are also usually paid a 50% premium or "for nothing" pay for the hours worked in excess of 40 in a week or 8 in a day, and a premium or "for nothing" pay of 5%, 10%, or 15% percent of their base wage rate for night shift work. In most industries, a 100% premium or "for nothing" pay is added for work performed on holidays and Sundays.

To establish a standardized basis for dealing with labor cost, this text will assume the conversion of all salaries into an hourly wage rate. We will treat all bonuses and incentives as a "for nothing" cost and will feed their cost into the product via the "use rate." This approach allows use of a standard or common basis for making labor and product/part cost calculations.

B. INCENTIVE COMPENSATION PLANS

In many industries, especially in labor intensive industries, some form of wage incentive is used to foster productive efforts. Of the basic incentive pay structures, the most common are:

1. *Standard hour incentives* are based on measured (time study) or synthesized (predetermined times) time standards. If the worker produces in excess of the standard, he is paid for the standard time allocated to his total production, even if it exceeds the base day's standard time (8 hours) production quota. This is the most common incentive plan. In this case, the company based fringe benefits do not vary, but the social security and other wage taxes vary with the amount of wages the employee earns. The overtime premium also applies in every case.
2. *Piecework incentives* usually pay a fixed price for each piece. They differ from standard hour methods in that the worker usually is paid only for the work produced. However, in most cases a minimum daily pay (call-in pay) is set. These incentives are most applicable in highly labor intensive operations like sewing and assembly. The same fringe and overtime rules apply in these wage systems as well.

Labor Cost

3. *Group bonus plans* are based on standard hour or piecework techniques but a group or team shares in the bonus. This technique tends to be self-policing since the group will prod laggards. The shares are usually calculated on the basis of the individual employee's wage rate.
4. *Shared gain plans* split the productivity gain above standard between the worker and management. Such plans are seldom used in today's labor environment.
5. *Profit sharing* is a distribution of a portion of the company's profit, usually on the criteria of worker wage rate or level, length of service, and profit performance. This technique is not common because of the difficulties arising from "open books" and disputes over the definition of "profits." Some public companies and private corporations do have supplementary profit sharing plans in which the payments are distributed in the form of an annual or semi-annual bonus.
6. *Negative incentives* are sometimes used for maintenance and setup people and supervisors. In such cases, a standard hour base is used to compute the average bonus for the department or work area. The supporting employee loses the bonus when the operation stops for repairs, setups, bad work, etc. The worker receives the average bonus when operations are normal.

For the purposes of this book we will assume that all workers and salaried personnel are paid on a standard hour basis, and that all employees in an enterprise receive the same fringe benefits. The management executives, supervisors, and other G & A personnel will be assumed to be on a fixed salary plus fringes with no overtime or night shift premium ("for nothing") pay.

C. MEASURED DAY WORK AND SALARIES

Beyond the need to record hourly workers' time on the job so that they can be properly paid, there are several reasons for recording attendance in an industrial plant. It is necessary to account for the employees' presence, absence, and time off for the management of sick pay days, vacations, service records, etc. In general, it is good business for management to know who is working and when. Labor records are also subject to audit by the State and Federal Labor Departments to assure proper payment of wages and compliance with wage and hour laws, social security regulations, and tax laws.

In most cases, salaried employees are not required to "punch in," and they are assumed to be on the job unless their absence is recorded with an approved reason such as illness or vacation.

In some industries, and as a part of some labor union contracts, the amount of work to be performed each day or hour is defined for each task and worker. The definition of the task is based on a "measured day" of work, by using a conventional work measurement procedure. In such cases the standard production rate is defined, usually by work measurement methods. However, the worker is paid for the full day regardless of production performance.

The worker's performance is evaluated in relation to the completion of the defined task. Raises and/or retention decisions are based on the employee's performance. Work schedules are based on the assumption that each worker will fulfill the defined "quota" in a measured day work system. Cost standards are also based on this assumption, or on a planned modification of it to accommodate failures to meet the schedule.

The basic difference between this payment policy and the standard hour system is that the worker is not paid for exceeding the schedule, and is not "docked" for failure to meet the quota. In the standard hour incentive program, the worker is paid for overproducing the schedule and is only guaranteed the minimum "call-in pay" if he fails to produce the scheduled quantity. In both cases, the "standard cost" of the task is clearly definable, and the over or underproduction constitutes a variance from standard.

In difficult to standardize industries such as computer building, software manufacturing, electrical and electronic machinery, and bioengineering, factory workers are paid on a salary basis and the work schedules are flexible. In such cases, the work standards or production elements to be used for standard costing can best be defined by work sampling and measuring job patterns rather than individual work tasks or cycles.

Regardless of the payment plan, peer pressure often has an impact on productivity. In well-managed plants with high morale, peer pressure tends to support production objectives and schedules are usually met and often exceeded. In other plants, particularly in union shops with adversarial relations between management and the union, peer pressure tends to limit the effectiveness of any wage incentive plans. In any case, the use of measured work standards provides a sound basis for a standard cost system with measurable variances.

Most of the salaried personnel in a manufacturing or warehousing enterprise are usually considered to be a part of management. Their labor cost is fed into the product cost as a part of burden or overhead via the "use rate." These personnel include the executives, supervisors, engineers, computer staff, and supporting administrative staff. In most instances, their productivity or task performance is not formally measured or standardized.

D. FRINGE BENEFITS

We have mentioned such fringe benefits as holiday pay and paid vacations, and such fringe costs as social security, worker's compensation, and unemployment

Labor Cost

compensation taxes. In today's industrial environment, these are only the beginnings of the fringe costs of labor.

Most manufacturing and warehousing enterprises provide some form of medical insurance, life insurance, and retirement pension plan for their employees. With the rising cost of medical care, these programs are increasingly significant as inducements to the hiring and retention of good employees. With the government entering the health insurance field, their cost will continue to rise in the forseeable future.

Some companies are also providing childcare for single parent or dual working parent employees. This is often in the form of a professionally staffed childcare facility. In most cases, the employee pays a nominal fee for the service, but it must, of course, be company subsidized. Federal law also requires granting of up to twelve weeks of unpaid leave for child and family home care. This law requires continuation of the employee's health insurance and pension program during this leave time, and the guarantee of the same, or equal job upon returning to work.

Many firms also underwrite the education of their employees through tuition reimbursement or scholarships. They also finance seminar attendance and run in-plant training programs. This educational support is usually different for hourly and salaried employee groups.

Another fringe benefit sometimes provided is the operation of an employee recreational establishment. These vary from baseball fields on plant sites to fully equipped gymnasiums and vacation resorts. Employees are often charged a user fee at the resort, but the facilities are usually heavily subsidized by the company. The cost impact of all of these, and other occasionally added fringes, generates a typical payroll fringe benefit factor of 35% to 50% of the base pay of the employee.

To develop a clearly defined labor cost base for a "total absorption standard cost," let us include all of the universally applicable individual employee benefit costs as a part of the labor cost. However, specialized benefits such as the educational subsidies, childcare services, leave time costs, and recreational facilities are treated as overhead items and their costs feed into the product via the "use rate" charge.

For the purposes of this text, assume that the universal personnel benefit cost is 35% of the base pay, not including the "for nothing" cost of overtime, holidays, and night shifts. In practice, this will vary with the particular company.

E. LABOR STANDARDS AND VARIANCES

As discussed earlier, the basis for an engineered costing system is the combination of a sound work measurement or work standards program with a "total absorption standard cost" procedure. The result of this combination of techniques is a set of "costed labor standards" for each operation on each part or assembly, or each handling task, in the enterprise. But let us first define what we mean by a "labor standard."

The first step in the design of a manufacturing operation is the development and definition of the process to be used in the production of the desired product. To accomplish this, we must break down the product into its manufacturing bill of materials (MBoM) and define the process for the manufacture of each part in the MBoM. The definition of the manufacturing process is usually documented in an "operation sheet" or "router" with supporting drawings and procedures. These can be either manual or computer prepared documents.

For each operation, the operation sheet or router specifies the kind of operation or task, the tools and machines to be used, the work center or location where the work is to be performed, and the time that the operation or task is expected to require. The time statement may be computed on the basis of the required machining time and its complete "floor to floor" part loading cycle, or, the operation time may be the expected worker time required to perform the operation.

In either case, the time prediction will usually be computed by the industrial engineer using predetermined time standards such as MTM or MOST plus the computation or simulation of the machine running time for the operation. When the job is later actually being performed, the predicted standard time can be verified or corrected by using a stopwatch study or work sampling. The corrected time standard will be entered into the operation sheet and it will provide the basis for work scheduling and costing.

One of the problems of modern manufacturing is the non-linear relationship between labor time and a machinery dominated manufacturing process. In many cases, the worker performs his tasks within the machine cycle and has idle time. In other cases, the worker may monitor or serve two or more machines. The machine cycle time is often the operation's controlling time and the basis for scheduling. The worker's time is then dominated by the machine cycle.

In other operations, the worker may dominate the work cycle and the machine may be idle during part of the time. This would be typical of an assembly or welding operation where the machine is only required during part of the work cycle. In this case the machine time is dominated by the worker's task cycle.

In any case, the worker is on the job for the full time and must be paid even when he or the machine is idle during part of the work cycle. The industrial engineer must also define the share of the worker's time which is to be assigned to each machine or work station in a multiple machine job. Because the machines may also be working on different parts or products in a multiple

Labor Cost

machine job, the worker's time and cost may need to be separately allocated to each item.

All of the worker's time must be assigned to the task in a single machine dominated cycle. The worker time definition is not a problem in a worker dominated cycle.

Thus it can be seen that the definition of the "standard labor time" for each operation will differ in accordance with the design of the operation. In some cases, the standard time for the worker will be the machine cycle time and in other cases the worker's time will be divided between machines or tasks. The establishment of the "labor standard" is based on the industrial engineering analysis of the task and the proper definition of the worker's role in the operation. The role must be defined in terms of time.

Labor standard variances can be caused by a variety of changes in the production situation. If the operation's process or procedure is changed, the labor standard must be changed at the same time. This will not be considered a variance. It is a methods change. This cost impact should be carried into the product budget and the inventory valuation.

However, if the process labor time per unit of production changes because of a machine break down or tool failure, a worker error or slower work pace, material failure or scheduling delays, this will cause an unfavorable labor time variance versus the standard labor time and cost. If a worker with a higher than usually assigned pay grade is performing the work, there will also be an unfavorable cost variance, even if the work time is not altered.

Conversely, better worker performance, the impact of an incentive program, or avoidance of expected rest and delay factors cause favorable variances in the labor time versus the units of production. In most cases, if a lower rated employee does the work, he will be paid at the normal wage rate and the operation cost will not be affected. The wage rate variance will show up as an increase in the payroll account.

In all cases, these variances will be computed for the individual operation and product or part in the WIP inventory. They will then be assigned and treated as positive or negative burden items. Their accumulated period (annual or six month) cost impact will later be allocated to the product standard cost via a periodic adjustment to the standard "use rate" charge. The direct labor charged to the product valuation will be computed from actual time and based on the standard labor cost.

F. MACHINE AND SYSTEM PACED OPERATIONS

In many industries, the pace of operations is largely dominated by the machinery. A good example of this is in a soft drink or beer bottling line. With line speeds of 700 to 1400 bottles or cans per minute, it is way beyond any human capability to do more than monitor the machinery. Workers are required to set up, monitor, clean, and troubleshoot the operation. When such machinery is running well, the worker is on a standby basis and cannot influence its performance in an appreciable way. In such cases, the cost of labor is a fixed expense, and must be assigned to the product unit cost on the basis of the standard operating rate of the machinery.

Production speed adjustments are treated as methods changes that change the standard labor charge. Labor cost variances can result from malfunctions in the line that generate a higher than expected unit labor cost. These variances are introduced into the product cost via the labor portion of the work station's "use rate" element in the actual cost computation.

As automation has expanded, more manufacturing operations have become machine paced. In all such cases, the required operator's standard labor cost should be assigned to the operation for the period of time that the machine is operating. When the machine is out of service, the operator's time cost should be assigned to other tasks that he may be given. If the worker is on a standby idle basis, the time cost is treated as an indirect expense and should be charged to the production cost via the "use rate."

G. LABOR REPORTING AND DATA COLLECTION

As pointed out in the previous sections, labor time and cost are not always linearly related to machine process time and work station production rates. As a result, the collection and reporting of labor time and cost data is a bit more complicated than might be assumed.

If the worker's cycle time and the machine time and/or production rate of the work station are parallel, the collection of labor time/cost data is simply a matter of recording the units of production and the total (or unit) labor time spent in the work station producing them. If the operation runs on a continuing or batch basis, the worker normally records or "clocks on" to the job order number and "clocks off" the job when reporting the number of good and bad pieces completed. This can be accomplished by the bar code wanding of the worker's badge, the job order number, and the task menu, or by keying in the quantity in a computerized system.

In a less sophisticated system, the worker might "clock on" and "off" with a time card or a telephone call to a time clerk or dispatcher. The piece count would be written on the job card or orally reported to the dispatcher.

Recording the units produced at each work station is a normal monitoring procedure for production and work-in-process. When the labor time is parallel to production, posting of the unit record automatically picks up the standard labor time along with the quantity of products. This procedure records the actual standard labor and machine process time. Excess or idle labor time is picked up from the worker's attendance report (wanding or punching in and out of the plant) minus the accrued standard labor time on completed production. This excess time is either treated as a labor variance on the assigned work station or allocated to a different task.

The actual time and production data is compared to the predicted standard time and cost for the operation to compute variances. The cost system uses the totally absorbed standard cost for the inventory valuation. The values of the variances from standard are recorded in an operations variance account within the burden accounts of the enterprise. The cost effect of these variances will be periodically applied to the "use rate" computation and fed into the cost of the product through the revised "use rate" as part of the burden cost.

When the labor and machine times are not parallel, the procedure is a bit more complicated. This is particularly true when the work station operator monitors more than one machine, and the machines perform different operations, often on different parts. In these cases, the standard labor time allocated to each task or unit of production must be defined by either work measurement or predetermined time standards.

Once the standard labor time per unit is established, the routine collection of actual unit production data automatically picks up and accrues the actual machine time and the assigned labor time used at the standard time rate. Using the total absorption, standard cost procedure, the unit cost can then be computed and assigned to the inventory valuation. The variances are defined by comparing the actual production to the planned output and the actual machine or work station time recorded by the worker's "clock on/clock off" procedure to the standard time. These data will be compared with the predicted unit or total time for the number of units of completed production and the on-the-job labor time of the operator.

The standard time/cost of the completed production will be posted to the inventory valuation (Figure 9.1). The excess time value of labor or "use rate," or any time savings values, will be treated as a variance. These variances will be posted to the actual cost column of the inventory to compute the operation or inventory value variance. The variance will be posted to the burden accounts for distribution via the "use rate."

In summary, the labor cost/time data are recorded as a part of the production reporting system and the worker attendance and/or job assignment reporting system. Any variance from predicted standard labor utilization is considered a variance and will be fed into the product cost as a periodic burden item via the "use rate" charge.

H. DEFINING LABOR COST STANDARDS

The first step in developing labor cost standards is establishing the labor time standards for each operation and for each part or product. This requires work

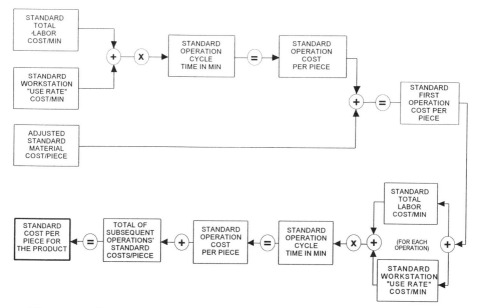

Figure 9.1 Standard part and product cost buildup concept.

measurement or the use of predetermined time standards to define the labor time content of each manufacturing, materials handling, and warehousing operation. This information must be documented on the operations sheet or router, or in the computer data base file's equivalent of the operations sheet. The next step requires definition of the prime cost of each hour or minute of labor.

Obviously the dollar value of labor time varies with the individual worker and the skill level or job description for the work being performed. However, with this variation in mind, one can define the cost of a labor hour. To cost an operation or product, one can apply the current dollar value of each specific labor hour used. In the case of management or salaried staff employees, the annual salary is converted into an hourly rate and the same fringe cost elements apply.

In a mathematical format, the manufacturing and warehouse labor cost (L) is based on the following:

$$L = W + T + F + N + B$$

where:

L = total labor cost or rate
W = basic employee wages or salaries (hourly rate)
T = legally required payroll taxes and insurance

F = management/union sponsored fringe benefits
N = "for nothing" costs based on schedules
B = incentive pay and bonuses for performance

In the proposed time use of facility, total absorption standard cost approach, the "for nothing" cost (N) and incentive pay (B) are treated as schedule driven overhead items. They are charged to a burden account and periodically distributed to the product cost through the "use rate." These labor cost items are not included in the direct and assignable indirect labor (L) rate. Thus, the primary labor cost rate is:

$$L = W + T + F$$

The standard labor cost factor (L_S) for each work station is the sum of the hourly direct process labor and the assignable hourly indirect labor which can be charged directly to any manufacturing operations at that work station. The standard hourly labor cost factor (L_S) for a work station is:

$$L_S = (L_D + L_I)$$

where:

L_S = the standard labor rate
L_D = the total of direct labor
L_I = operation assignable indirect labor for that work station

Calculation of the standard labor rate includes the basic wage rates, the labor taxes (workers compensation, unemployment compensation, and social security), and the hourly cost of such fringe benefits as health care and life insurance, vacations, and holidays. Thus, each work station has a specific standard labor rate which will be adjusted annually or periodically to accommodate wage rate and fringe benefit changes.

Non-linear labor cost items such as educational benefits, child care costs, family leave costs, incentive pay, and "for nothing" pay are treated as overhead or burden. They find their way into the product cost through the "use rate" charges.

To define the standard labor cost (L_S) for each part, operation, and work station, it is essential to have an in-place work measurement or predetermined standard cost system. The definition of the standard labor time for each operation should be a measured or computed time value in minutes or decimal parts of minutes. The value of a minute will vary with the wage rate of the assigned employee and the impact of fringe benefits. The dollar value of the minute should not be applied until the time standard is firmly defined and applied to the specific operation.

The primary labor charge (L) for each operation and part is calculated by multiplying the "floor to floor" operation cycle time (t) in decimal minutes per piece by the standard labor cost rate (L_S) to compute the labor cost per piece (L) for the part and operation.

$$L = L_S \times t$$

Burden labor (L_B), manufacturing management labor (L_M), and indirect labor (L_I) which is not specific to an operation are computed in the same manner. However, they will not be assigned to an operation or part. They will be included in the calculation of the costs which are a part of the work station's "use rate." This will be discussed later.

I. GETTING LABOR COST INTO THE PRODUCT BUDGET

As stated earlier, the primary vehicle for the accumulation of costs into a product budget is the work-in-process inventory record. All manufacturing and warehousing costs for labor, materials, burden, and overhead must flow into the inventory record for the valuation of finished goods. The only means of recovering these costs is through the sale of the inventories.

As each operation is completed, the actual time consumed is recorded in the computer data base as a part of the timekeeping and production management system. The completion of the task also results in the recording or posting of the quantity of pieces or units completed (and scrapped or reworked) into the inventory record. This data collection process can be performed by manual documentation, timecard and timeclock procedures, or by using bar code readers and key pads for direct entry into the computer.

At the same time, the computer recovers the standard labor cost factor (L_S) and the standard labor time (t_S) for the operation from the data base. Concurrently, the actual time (t) consumed in the operation is multiplied by the standard labor cost factor and the resulting actual unit and batch cost is entered into the inventory data base record. Any variance between actual and standard is calculated and posted to the inventory data base for comparison. The resulting computed variance is also posted to the variance file of the burden data base for periodic inclusion in the evaluation of the work station's performance and "use rate" computation. It will also be reflected in the manufacturing enterprise's profit and loss calculations.

The product's budget is computed from both the standard labor cost and the actual labor cost by retrieving these data from the inventory file along with the value of the materials used and the accumulated "use rate" charges. The work-in-process inventory is valued on a cumulative basis at each operation or step in the process. The introduction of the material and "use rate" cost elements into the inventory valuation process will be discussed later.

10
Operating Equipment Cost

A. CAPITAL EQUIPMENT INVESTMENT AND BOOK VALUE

With the increasing capital intensity of industry, the impact of equipment ownership cost is becoming more of a factor in the computation of product costs. An enterprise's investment in capital equipment is the total of the "landed cost" of all of the equipment in the plant. The landed or in-place cost is the price of the equipment plus the cost of its delivery, permanent tooling, and installation.

The "book value" of the capital equipment shown in the financial statements of most enterprises is normally based on its depreciated value, not its purchase price or replacement value. The book values of capital equipment vary with age, the initial cost of the asset, the choice of depreciation method, and the allowable tax life of the equipment. The owner may have the option of using straight line, double declining balance, or sum of the years digits methods of depreciation, or be required to use the seven year MACRS method approved by the Internal Revenue Service. Each depreciation method results in a different annual depreciation rate. The rapid write-off methods result in a different cost of ownership and book value for each successive year.

In dealing with this issue, it should be clearly understood that there are major differences between the purposes of operations cost accounting and financial accounting. As stated earlier, if the financial manager's goal is to present a picture of the enterprise's operations to bankers, credit rating agencies, owners, stock holders, the Internal Revenue Service, and possible purchasers of the business or its shares, the accountant usually uses "generally accepted accounting principles and practices." These practices are based on distribution type accounting. They use a variety of rapid write-off depreciation techniques to maximize the tax saving impact from depreciation of assets.

These distribution accounting techniques usually result in a variable and inaccurate definition of product cost because they tie the product cost to the volume of production, variable depreciation schedules, and such variable bases as labor and material cost.

In contrast, the stated objective of engineered precision costing is to identify, capture, and define the true cost of manufacturing and handling a product through the production and distribution system by using a costing method which is independent of taxation considerations, production volume, and the scheduling of operations.

A basic part of the total absorption standard cost accounting philosophy and the time use of facilities or "use rate" approach to product costing is the manner in which it deals with the capital value of fixed assets and the cost of their use in the manufacturing and warehousing operation. In developing the "use rate," the equipment and other fixed assets are valued at their initial landed costs and depreciated on a straight line basis for the IRS allowable life unless expected technical obsolescence dictates a shorter life. If the equipment is specifically required for the manufacture of a product with a predictably short market life, the depreciation life may also be shortened.

In this precision cost system, the current book value is not used in the calculation of the overhead or burden cost. The equipment usage or ownership cost calculation is based on the straight line depreciation as charged against the initial landed cost of the asset. The straight line depreciation charge is treated in the same manner as a rental or lease rate expense. If the equipment or building is leased, the lease charge is substituted for the depreciation rate.

B. DEPRECIATION SCHEDULES AND TAX LAW

The use of a straight line depreciation schedule for the IRS life of manufacturing, materials handling, and computer equipment or real estate is acceptable to many accountants. However, the tax accountant has a different agenda from the operating executive. Tax law also specifies the acceptable depreciation methods for use in the calculation of income taxes.

The operations executive needs a firm base from which to measure variances and trends in the cost of manufacturing and warehousing operations. The operations executive does not want the apparent cost of production to vary from period to period because of variable depreciation charges or swings in production volume. He wants to see the cost variances which result from changes in the product, the process, or an improvement or drop in the productivity of the plant.

Operating Equipment Cost

The use of straight line depreciation schedules to accumulate the ownership cost of assets into the product cost stabilizes one of the accounting variables used in the calculation of the "use rate." The time use of facilities concept of total absorption standard costing treats the straight line depreciation charge, and its associated interest cost, as "rent" for the actual useful life of the equipment or asset, even if it exceeds the IRS tax life. This depreciation or "rent" charge is included in the equipment "use rate" and the building and work station's space "use rate" (Figure 10.1).

C. MAINTENANCE, MODERNIZATION, AND REPLACEMENT COSTS

Machinery and equipment wear out! This creates a maintenance cost (m_E). Machinery and equipment also have a capital value. This requires depreciation. There is also an interest charge on the investment, and, advancing technology obsoletes manufacturing equipment and generates a replacement cost.

These costs are generally non-linear manufacturing expenses. They are theoretically covered by the depreciation's sinking fund, but that is not necessarily true. A separate capital equipment account is required to track these expenses. This account is an element of factory burden and should be charged into the product cost through the "use rate." To apply these expenses to the "use rate," they must be converted into an hourly charge.

In the case of fixed work station equipment, this charge can be based on the standard time procedures discussed above. In the case of forklift trucks and other mobile equipment, time use can sometimes be more effectively defined by measuring the fuel or energy consumed per hour and monitoring the fuel or energy usage to define the hours of use.

To establish a proper capital account, each machine should have two identifying numbers. One of the numbers is the permanent capital account identification of the machine or equipment, and the other is its current location and operation assignment code. These numbers can be combined as shown in Figure 7.17. The basic capital account number leads to the capital account data base which includes the identity of the manufacturer of the equipment, the equipment model number, the date of the unit's acquisition, the manufacturer's serial number, the initial purchase price and landed cost, and the current location of the equipment.

Alternatively, the work station number can use the SIMSCODER bill of material format. In this case, the indent number would be the same as the level of the work station's position in the process. For example, finished part machining equipment would be assigned a finished part indent code, assembly station equipment would be assigned the assembly indent, and finishing equipment would be assigned the painting or plating indent. The function field would

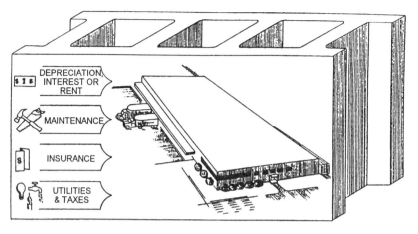

FACILITIES COST
(USE RATE)

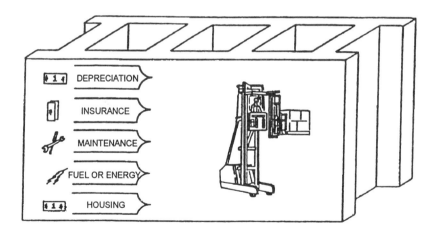

HANDLING EQUIPMENT COST
(USE RATE)

Figure 10.1 The concept of costs in the building and equipment "use rate."

Operating Equipment Cost

apply the same code translation for the type of equipment as in the permanent equipment identity code (i.e., lathe, mill, press, drill, etc.). The random number of the code would be the department or cost center code to tie the equipment unit to the financial accounting system.

The capital equipment account number can use the SIMSCODER type format in a slightly different format as shown in Figure 7.17. In this arrangement, the product field will identify the plant location of the equipment, the indent field shows the profit center or department where the equipment is currently located, and the function field defines the machine type or class of equipment (i.e., lathe, mill, drill, computer, press, etc.), and/or work center function. The random number field shows the current machine number in the department where it is located. This department/machine number can also be used as the work station number for shop loading and preparation of operations sheets. This number should be stenciled or painted on the machine or on a work station identity sign.

The item number field can be a random identity code or it can be sequenced to the machine number in the department. It could also be structured to include the last two digits of the year of acquisition and the sequence number of machines purchased or moved into the department in that year (i.e., 94XX for up to 99 machines purchased in 1994). This number will change when the machine is moved to another department.

The capital account number in the other portion of the code is permanent. It should be stamped into a metal tag and attached to the machine near its manufacturer's name plate which will usually show the machine's model and serial numbers. The capital account number will show the year of purchase and a purchase sequence number for that year. In situations which exceed 99 departmental acquisitions or 9999 units of purchased capital equipment in a year, the four digits can be the sequence number within the year. A "variation code" leader can show the year of purchase.

The current number assigned to the equipment should be its current work station number. This number is used in the writing of process or operations sheets or routings. It defines the location of an operation for planning, scheduling, and reporting work progress.

Both of these numbers should be included in the capital equipment data base or on the ledger card in a non-computer system. These numbers are the tools or "buckets" and "blocks" (Figure 10.1) for capturing and recording equipment maintenance, operating, and ownership costs and the allocation of these costs to burden accounts and the work station "use rates." Period costs, such as depreciation and interest, are charged to the capital account by conventional accounting transactions. Equipment maintenance charges are collected into the capital account by a maintenance work order system. Initial equipment purchase costs are entered through the purchase order procedure.

In charging maintenance and modernization to the capital equipment account, it will be necessary to separate expense charges from capital investments. In general, modernization or upgrading of the equipment, if the cost is significant, will be considered a capital investment and an acceptable IRS approved depreciation charge must be assigned. For costing purposes, this will be a straight line rate. Conversely, maintenance and repair work is usually treated as a current expense.

The total cost of any equipment upgrade will be capitalized. In defining the "use rate" value, the depreciation charge for the modernization is added to the equipment's basic depreciation charge. These combined depreciation charges, plus equipment operating expenses, are used to calculate the new "use rate." The modernization investment is treated as a part of the invested capital in the computation of the G & A charge. This will be explained later.

To collect maintenance expenses, each piece of equipment, and each building, will have a maintenance "account" under its capital account identity or tag number. In a manual system this record would be in the form of a ledger card. In a computer it would be in a data base file. In both cases the equipment "tag" number or building number would tie the capital item to the cost record.

The media for accumulation of maintenance costs into the data base or maintenance ledger are the maintenance work order and the replacement parts purchase orders and stock room requisitions. The work order document is tied to the equipment through the capital account or tag number and the work station or current machine number. The purchase order cost is tied to the work order by a store room requisition to withdraw the maintenance supplies from inventory.

The use of the work station number allows accounting to analyze the cost of maintenance by machine and department, and even by worker, through the operator's clock number. It also allows the assignment of maintenance costs to the "use rate" for each work station. The data base is also the source of maintenance history for each piece of capital equipment and for buildings.

All maintenance labor and material charges are gathered into the work order. The total cost of the work order is charged to the capital account through the equipment's tag or capital account number. The maintenance work order should include (or have computer access to) the tag number, work center number, work order number, issue date, work task description, work start date and time, work completion date and time, and hourly labor and "use rate" charge rates.

The sum of all equipment maintenance charges and work center costs is accumulated into the "use rate" for the work center. Building maintenance costs are charged to the product through the work center's space "use rate."

D. OPERATIONAL SUPPORT SYSTEMS AND UTILITIES COSTS

Manufacturing and materials handling equipment requires support systems, fuel, and utilities to operate. These include compressed air, water, electricity, natural gas or propane, and diesel or gasoline fuel.

The use of some of these items can be metered, measured, or computed for each piece of equipment. For example, the electric power used by a machine can be either metered or computed on the basis of motor sizes and other power consuming features. Combustion and motor fuels can also be well-defined on an individual equipment usage basis. Water can also be metered for each user or computed on the basis of pipe sizes and pressures. This is not as easily done for compressed air but the cost of the compressor can be distributed through the "use rate."

This expense is in addition to the fuel and energy used for environmental management of the plant as a whole. The environmental energy for lighting, heating, and air conditioning can be considered a part of the operating cost of the building. Its cost should be distributed to the work stations through the space utilization "use rate" procedure.

All of these supporting expenses are factory burden items and must be included in the work station and/or space "use rate" charge.

E. TOOLING AND OPERATIONAL SUPPLIES COSTS

The operation of manufacturing equipment also requires the use of durable and consumable tooling and production supplies or "contingent" materials. Let us first define these categories.

Fixed tooling includes the specially designed jigs, fixtures, dies, patterns, and molds which are specific to a part or assembly and remain in the system as long as that part or assembly is being manufactured. These tooling items can be capitalized and amortized on the basis of their expected time life or on the basis of the number of pieces to be produced on them. The units of production approach is most appropriate when a specific product quantity or product market life is definable or the tool's service life in terms of pieces produced is predictable. These fixed tools should be treated in the same manner as machinery and their amortization expense should be converted into a "use rate" charge for the work stations that make use of them.

Consumable tooling such as tool bits, drills, milling cutters, saws, files, etc., are a different matter. These items, and such "contingent" materials as emery cloth, abrasives, cutting oil, lubricants, wipe rags, work gloves, solvents, etc., which do not become part of the product, should be treated as manufacturing burden materials. The expense of these burden items should be distributed to the product through the vehicle of the work station "use rate."

In situations where usage is confined to a specific department or work station, the "use rate" for these burden items should be applied to that location only. In any case, these consumable tools and "contingent" materials must be charged to the product and operation through the work station's hourly "use rate."

F. UTILIZATION FACTORS AND IDLE TIME COSTS

As stated earlier, a key philosophical concept in the time use of facilities approach to precision engineered costing is the idea that the product being made in an operation or work station should not be charged with the unused idle time of the balance of the manufacturing facility or the work station. It should only be charged with the time use cost of the equipment and space that it actually uses.

The concept of the "use rate" is the same as the philosophical basis for renting an apartment or hotel room. The tenant only pays for his apartment or room for the time he is occupying it. The empty apartments or rooms are supported by the management and their cost impact is on the profit and loss statement, not on the tenants. One might argue that the rental rate includes the cost of some portion of the expected empty space or idle time. This is possibly true, but most of the empty space is a cost charged to the enterprise and not the tenant.

In structuring the "use rate" or rental charge for a work station, this same philosophy must be applied. The expected time use of the work station or facility must be defined on the basis of capacity needs, work schedules, and workloads. This time use will provide the basis for defining the value of the "use rate."

If the machine or work station is expected to be fully loaded, the one shift usage would be 2080 hours per year minus any vacation and holiday shut downs and a reasonable expectation of operating down time. This might result in a planned operating year of 1952 hours on one shift with two weeks of vacation shut down and six holidays. In addition, one might assume 10% down time for unscheduled maintenance for a total availability of 1756.8 or 1757 operating hours per shift year. This would be the number used to define the hourly "use rate" charges for that machine or work station. All burden, overhead, assigned labor, and space costs in the "use rate" would be computed on the basis of 1757 hours per year for a full shift of work on a 40 hour week. Figure 10.2 shows the method of calculating the "use rate" for power materials handling equipment and Figure 10.3 shows the same procedure for manufacturing equipment.

But what do we do about work stations that are only used part of the time? What about broaches, heat treating furnaces, burring systems, shotblast rooms, etc., which are used intermittently? In these cases, the expected scheduled time (perhaps half or one third of the full shift) will be used as the basis for calculating the "use rate" for that work station. The cost of the scheduled idle time will be absorbed by the "use rate" for the operating time.

However, in the case of unscheduled or uncommitted idle time caused by lack of work, the cost of the "use rate" charge for the idle time will be treated as a variance and an adjustment to the profit and loss statement. It will not be charged to the product or operation. It will be charged to the general expense account of the enterprise.

For example, if a work station is scheduled to operate on a full time basis, and if it is only needed for half of the time because of a downturn in business, the "use rate" charge for the idle time will not be charged to the product being made at that work station. It will be charged to the enterprise as an expense item and will impact the profit and loss statement. Only that portion of the time used to produce the product will be charged to the product's manufacturing cost. The product will "rent" the work station for the time needed for its manufacture, and no more!

Therefore, the "use rate" charge for any equipment, work station, or space will be computed on the basis of its scheduled usage and not the potential for utilization. Only that portion of the available time which is utilized will be charged to the product. Idle capacity cost will be charged to the enterprise's profit and loss statement as a miscellaneous overhead expense.

G. EQUIPMENT UTILIZATION COST STANDARDS—"USE RATES"

In order to establish the "use rate" for the manufacturing and materials handling equipment, it is first necessary to define the applicable annual cost elements for each equipment item. When these cost elements have been defined, the appropriate scheduled annual hours of operation must then be established for each unit or work station. As pointed out above, the scheduled hours will vary from full time to intermittent use. A full time schedule for a 40 hour week usually results in only 1757 hours of actual production per year. On a full time schedule, this could be the appropriate number of hours per shift to use in calculating the hourly "use rate" from the annual cost.

To define the annual cost of the equipment, all elements of cost must be included. In a mathematical format, the annual cost of equipment would be:

$$E_A = D\%(E_{pc}) + i\%(E_{pc}) + T_{pp} + I_c + m_E + U_E$$

where:

E_A = the annual cost of owning and operating the equipment
$D\%$ = the annual straight line depreciation percentage rate for the IRS allowable life of the equipment
E_{pc} = the initial purchase price or landed cost of the equipment

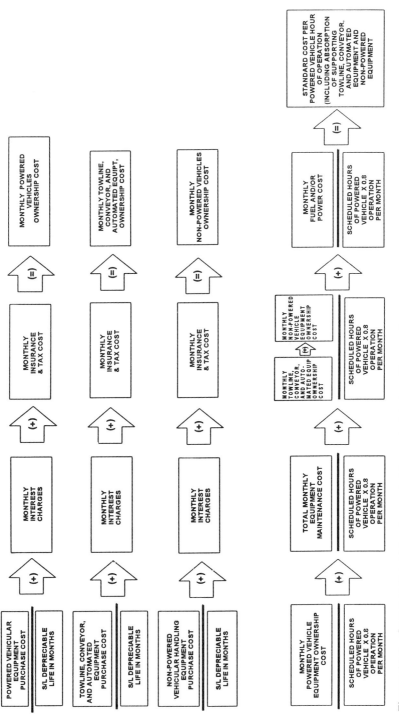

Figure 10.2 Computation of the "use rate" for materials handling equipment.

Operating Equipment Cost

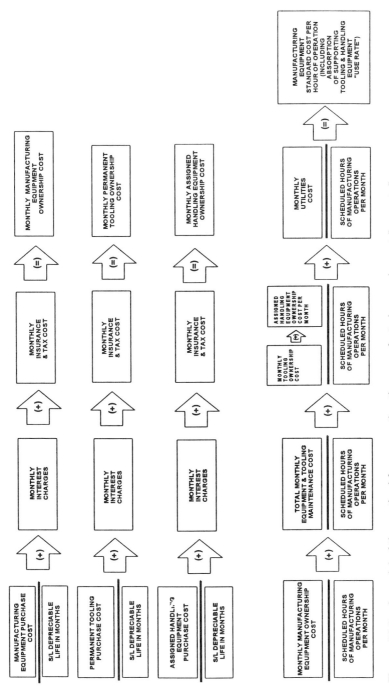

Figure 10.3 A model of the method for computing the manufacturing equipment "use rate".

$i\%$ = the annual interest rate, at the time of purchase, on the money spent to purchase the equipment

T_{pp} = the predicted annual personal property tax on the predicted book value of the equipment in the middle year of its expected service life

I_c = the annual casualty insurance cost (fire, theft, etc.) on the equipment

m_E = the annual expected equipment maintenance cost

U_E = the predicted annual fuel, electricity, compressed air, water, and other utilities to support the equipment

Using this formula for the annual cost of the equipment, one can compute the equipment's "use rate" by dividing the annual cost by the scheduled annual hours of operation, whether full time or intermittent. For example, based on the annual cost calculation shown above, the "use rate" would be:

$$E_A/1757 = \text{Hourly "Use Rate"} (E_{At})$$

for all equipment on a one shift, 40 hour schedule.

Figure 10.2 is a graphic model of the development of an equipment "use rate" for materials handling equipment. The same concept is applicable to manufacturing, office, and storage equipment with appropriate additions and omissions to fit the situation.

Figure 10.3 is a graphic model of the computation of the "use rate" for manufacturing equipment.

The same calculation procedure and formula should be used to compute the "use rate" for each individual item of equipment. The specific and applicable "use rate" will be used for each operation and work station used in a manufacturing plant or warehouse. The "use rate" will be stable as long as the components of the calculation remain the same. This standard should be reviewed at least annually for required adjustment in an operating system.

When the scheduled hours of operation are intermittent and/or less than a scheduled full shift, the same formula will be used with appropriately reduced annual hours of operation. A similar calculation can be used to compute the "use rate" for permanent or fixed tools. The "use rate" for the fixed required tools and the expenses for assignable contingent materials items should be added to each work station's equipment "use rate."

H. GETTING THE EQUIPMENT "USE RATE" INTO THE PRODUCT BUDGET

When the equipment "use rate" has been defined, it will be added to the space "use rate" charge to become the basic work station "use rate." The space "use

Operating Equipment Cost

rate" charge will include the general building and management overhead, the operating burden costs, and the general and administrative expense charges allocated to the work station. The method of calculating these space "use rate" cost elements will be discussed in the following sections and chapters.

When the complete work station "use rate" has been computed on an hourly basis, it will be divided by 60 to define the "use rate" per minute. Each operation's cycle time in minutes (t) is multiplied by the work station's "use rate" per minute to compute the "use rate" charge for the operation. The operation's "use rate" charge is added to the product of labor cost per minute times the operation cycle time (t) to compute the cost of the operation for the part or workpiece. In the first operation, the standard material cost will be added to the "use rate" charge and labor cost to compute the full cost of the operation. The sum of the costs of all operations is entered into the valuation of the inventory to produce the product budget. This will be discussed in more detail later.

11

Manufacturing/Warehousing Operating Burden or Overhead

A. MANUFACTURING/WAREHOUSING MANAGEMENT EXPENSE

Several cost assignment dilemmas are involved in the definition of management overhead expense at the operating level. One dilemma requires the definition of operations management costs versus general and administrative management expenses. Another results from the occasional unionization of line supervisors in some industries. A third dilemma involves the roles of the industrial and manufacturing engineers, and the middle management staffs in production and inventory control, personnel management, safety and security, and other supporting services.

In most cases, the accountant or controller makes the decision concerning proper assignment of the cost of these people and their work stations and supplies. Accountants also sometimes establish cost centers to segregate expenses by department, plant, function, or geographical areas within a plant. For the purposes of this book, and as a guide to those who will implement the proposed cost procedures, we offer some ground rules for allocation of these management and staff costs.

The following position statement defines the book's posture regarding the division between manufacturing/warehousing operations management and general and administrative (G & A) expense.

1. If the senior management of an enterprise includes a vice president (or in the case of European organizations, a director) who is specifically charged

with the management of manufacturing only (as in the case of a Vice President of Manufacturing), this executive should be considered to be a part of the manufacturing organization and not a part of the G & A expense or enterprise overhead.

If a senior executive is in charge of operations and his scope of responsibilities includes other functions in addition to manufacturing, he should be considered to be a part of the general and administrative staff and the overall enterprise overhead.

In either case, these executive costs will be treated as factory overhead and will be charged to the product through the "use rate," but the procedures for computing their impact on the "use rate" will differ.

2. If the senior management includes a vice president (or in the case of European organizations, a director) who is specifically charged with the management of warehousing and transportation only, or physical distribution only, as in the case of a Vice President of Distribution, this executive will be considered to be a part of the warehousing or physical distribution organization and its overhead, not a part of G & A.

If this executive is in charge of "operations" and his scope of responsibilities includes other functions in addition to warehousing, distribution, and transportation, he will be considered to be a part of the general and administrative staff and the overall enterprise overhead. All of these costs will be treated as overhead and will be charged to the product through the "use rate."

3. All plant managers, warehouse managers, superintendents, foremen, and supervisors will be considered a part of manufacturing or warehouse operations management in spite of the fact that some supervisors may be members of the union bargaining unit. These labor costs will be treated as factory or warehouse burden.

4. Working "leaders" or "straw bosses" who perform hands-on production functions in addition to being team or group leaders or supervisors will be considered to be part of the "hourly" work force. Their supervisory time will be separated and treated as factory or warehouse burden through the development of work standards for their jobs.

5. All supervisory and manufacturing and warehousing operations management costs will be charged to the product as an operations burden through the space "use rate" charge. The directly assigned vice presidents will be considered overhead and their their cost will also be charged to the product through the space "use rate."

6. In every manufacturing and physical distribution operation there is a cadre of personnel who are a significant staff support to the enterprise but have no clearly defined line management role. These include the production and inventory control managers and their staffs, the personnel manager and his supporting personnel, safety and security people, and in most cases, manufac-

turing, tooling, and industrial engineers who provide technical support to the operation. All of these personnel and their supporting facilities should be treated as a manufacturing and/or warehousing management burden cost. Their costs should be allocated to the product through the space "use rate" charge.

To permit more precise control of the manufacturing process, these management expenses may be segregated into "cost centers" or departments. As a result, the management burden "use rate" might be separately calculated for each area, cost center, or department, and may differ between departments. This is an acceptable refinement in the cost accounting procedure.

In every case, the management burden and/or overhead "use rate" will be allocated to the product cost on a time use basis. The calculation of the hourly "use rate" charge will be based on division of the total management burden and/or overhead for the plant (or cost center) by the scheduled annual hours of normal production. This will result in an hourly "use rate" for the plant (or cost center). This quotient will then be divided by the net productive area of the plant (or cost center) to derive an hourly "use rate" per square foot of net productive space. Multiplication of this figure by the assigned area of the work station will produce the management burden "use rate" for the work station. A similar procedure will be used for overhead assignment.

B. INDIRECT MANUFACTURING/WAREHOUSING LABOR

Another area of controversial interpretation is in the definition of indirect labor. In manufacturing, the definition of direct labor is usually based on the "hands-on" concept wherein the worker directly participates in the manufacture of the product. The fuzzy part of this definition involves the hands-on inspection and materials handling functions which do not change the condition or status of the part or product, but are a required part of the work flow. It is therefore necessary to again state some ground rules for the design of a system. These rules can, of course be modified by users of the system if the changes will fit their circumstances better. In this book we will use the following definitions for direct and indirect labor.

1. In manufacturing, the definition of *direct labor* (L_D) is confined to those personnel who are involved in, and have a hands-on relationship with the alteration of the shape, appearance, function, condition, assembly, or packaging of the product and its components. The definition of "hands-on" includes the manual, computer, and/or remote control of machinery, the loading and unloading of materials or parts into machinery, the assembly and/or packaging of the product, and the direct and continuous materials handling support of a manufacturing operation.

2. In manufacturing, the definition of *indirect labor* (L_I), as opposed to burden labor, is that it is the labor which is used to directly support manufacturing without changing the character or condition of the product. Examples of this class of labor would include product and process inspection, general plant materials handling, storekeeping, machinery maintenance, tool making and tool sharpening, setup labor, and scrap handling.

Machinery maintenance and tool making labor costs should be charged to the product through the equipment "use rate" procedure as described above. Other indirect labor costs should be included in the factory burden element of the "use rate." Whenever possible, the indirect labor should be assigned to a specific work station and be included in the "use rate" for that work station. However, when indirect labor supports the whole plant, or is not clearly identified with a particular cost center or work station, it will be included in the computation of the space "use rate" for the net productive space. The indirect manufacturing labor "use rate" charge will be assigned to work stations on the basis of their space utilization.

3. *Warehousing direct labor* can be defined as the labor which actually handles the goods. This would include receiving, transport to storage, placement in storage, order picking, transport to shipping, checking and packing, and outbound vehicle loading. These functions are treated in the same manner as the direct labor in the manufacturing operation. The standard time (t_S) is multiplied by the standard labor cost (L_S) for the operation and the result is compared to the actual time and cost to generate a variance if they do not match. The standard cost is applied to the finished goods inventory valuation and/or ownership cost.

4. *Indirect labor* in warehousing would include such operations as clerical support, housekeeping, equipment maintenance, inventory counting, and supervision. These costs are applied through the "use rate" in the same manner used for the indirect manufacturing labor and the manufacturing management expense.

C. FACTORY/WAREHOUSE OVERHEAD

When defining factory and warehouse overhead, we again find a difference in opinion concerning the proper allocation of some items. There are two levels of personnel and administrative overhead and overhead expenses in the system. One is the enterprise's management staff overhead. This is defined as general and administrative expense in this text. Another overhead expense is found at the factory and warehouse level when senior management is directly involved in the operation. This management labor cost can be combined with the manufacturing and warehouse burden in computing the "use rates."

Manufacturing/Warehousing Operating Burden

In developing the time use of facilities and total absorption standard cost system, we have made some assumptions and policy decisions to build the system. As long as the basic concepts are followed in its application, the system can be installed with variations in these assumptions to suit the local conditions.

In order to define "overhead" at the factory and warehouse level and to distinguish it from burden and indirect labor, let us use the following:

> *Overhead* at the factory and warehouse level is defined as the cost of supporting personnel and services which will be in place regardless of the volume or nature of the activity in the plant, and which are non-linear in their relationship to the productive activity. These costs would include the middle management salaries and their supporting secretarial and office administrative services.

Presented in mathematical format, this would be:

$$\frac{(F_{Bp} + W_{Bp} + F_{OHp} + W_{OHp})}{\text{Scheduled normal hours } (t_s)} = \text{Hourly plant "use rate" } (U_{Pp})$$

where:

U_{Pp} = plant burden and overhead personnel "use rate"
F_{Bp} = factory personnel burden
W_{Bp} = warehouse personnel burden
F_{OHp} = factory personnel overhead
W_{OHp} = warehouse personnel overhead
t_s = standard hours of operation

In order to apply this plantwide "use rate" cost to each work station, it must be converted into a "use rate" charge per square foot per minute. This figure can then be multiplied by the area of the work station to get a specific "use rate" charge for each minute of the work station's use for each operation. This charge would then be multiplied by the minutes required (or the standard time) for the operation to compute the total "use rate" charge for burden and overhead labor for the operation. In a mathematical format, this calculation would be as follows:

$U_{Pp}/60$ = "Use rate" per minute (U_{Ppt}) for plant burden and overhead personnel

(U_{Ppt})/net production space in square feet = "Use Rate"/min/sq ft(U_{Pptf})

$U_{Pptf} \times$ area assigned to work station in sq ft = U_{Ptws}

or:

> The "use rate" charge per minute for plant burden and overhead personnel for the particular work station

This "use rate" (U_{Ptws}) is multiplied by the number of actual and standard minutes used in the operation to produce the standard and actual plant burden and overhead personnel cost for the particular operation and part or item. The result will be included in the cost valuation of the work-in-process inventory or finished goods produced by that operation. The procedure will be explained later.

D. NON-PRODUCTION MATERIALS AND SUPPLIES

Although computers have eliminated much of the paper work from manufacturing and warehousing operations, many non-production or burden materials and supplies are still part of the operation. These include forms, computer supplies, and documentation materials. Other non-production materials are sanitary supplies, pencils, paper clips, sealing tape, stationery, rubber bands, photocopy materials, etc. All of these materials must be accounted for in the cost system.

The most practical method for absorbing the cost of these non-production supplies into the product cost is to accumulate their costs on an annual basis and distribute them through the space "use rate." By dividing the total annual cost of these supplies by the number of standard minutes in a production year and then dividing the quotient by the net operating space, a cost per minute, per square foot can be defined and included in the work station "use rate." These costs can be assigned to work stations through the "use rate" and the time use of their occupied space. Any unabsorbed cost of non-production materials should be charged to the profit and loss statement as a miscellaneous expense.

E. CONTINGENT MANUFACTURING/WAREHOUSING MATERIALS

As stated above, some of the cost items which are included in the manufacturing and warehousing burden can be defined as contingent materials. These are the consumable tools, coolants, wipe rags, emery paper, lubricants, foundry sand, documents, and paper products which are used in production, but do not become a part of the product. Some productive materials, such as welding rods, soldering rods, and plating electrolite, can also be treated as contingent materials. Their consumption is also non-linear in its relation to production, but required in the manufacturing process.

Most of these materials do not become a part of the product; they are used in a non-linear ratio to the product production volume. They sometimes relate to a particular cost center or department, but are seldom on a linear consumption basis in a work station. It is therefore logical to assign their cost to production on a shared consumption basis. A share of their accumulated annual cost should be charged to the products through the work station's "use rate" on a time use of facility basis.

The total annual cost of contingent materials should be divided by the standard scheduled plant operating hours. The quotient will be divided by the net plant operating space and the result will be divided by 60 to generate a consumption or "use rate" for contingent materials per square foot, per minute, for the plant. However, when these materials are used only in a specific department or work station, their cost should be assigned to that cost center's space "use rate" and not to the whole plant. In each case, this cost distribution figure will be multiplied by the square feet of space assigned to each work station as the means for allocating the cost of the contingent materials to the work station on a time use of the facility basis.

The results of this calculation will be added to the other cost elements in the work station's "use rate" to build the standard "use rate" for the work station.

F. INSURANCE, INTEREST, AND TAXES

In most conventional accounting systems, the costs of insurance, interest, and taxes are included in the general and administrative expense accounts of the enterprise. This is an appropriate allocation of these expenses. However, they are also treated as a "below the line" cost item on the profit and loss statement. This is unfortunate as it tends to imply an erroneously lower factory cost for the product.

In the total absorption standard cost approach presented in this text, the G & A is treated as an "above the line" cost. It is proportionately absorbed into the inventory valuation at each step in the production process. It becomes a part of the "factory cost of sales" of the product. The method of accomplishing this will be discussed in detail in Chapter 13.

G. BURDEN/OVERHEAD COST STANDARDS

As we can see from the foregoing discussion, there are a number of burden and overhead cost components which must be accumulated into the product's inventory valuation. The total absorption, standard cost approach and the time use of facility method of assigning costs to the product require development of a standard cost module for assignment of burden and overhead expense items.

Most burden and overhead cost elements are, or can be treated as, period type expenses. It is therefore logical to distribute these expenses to the product through the work station's "use rate." The "use rate" can collect all of the burden and overhead costs into a time and space based standard. This standard can then be applied to the product on a time use of facility basis.

The standard burden and overhead cost module will have a clearly definable fixed base. The standard cost will be assigned to the product cost or inventory valuation on the basis of the time use (in minutes or fractions of minutes) of the work station when performing the production operation on the workpiece.

The content of this "use rate" has only been partially developed in the foregoing chapters. The building and equipment or facility overhead, and the general and administrative expenses must be included to generate a complete work station "use rate."

The complete work station "use rate" must include the cost of all executive, supervisory, and indirect support personnel (U_{Pp}), non-productive indirect and contingent materials costs (M_{ic}), the cost of owning and operating production equipment (E_{At}), the work station's share of the G & A expenses (G&AX), and the work station's share of the building ownership cost (B_A). Application of the "use rate" standard to each operation will be based on the time use of the equipment in the work station and the space that the work station occupies.

When the G & A and building burden are included in the "use rate," it provides a "standard cost" for all of the indirect and/or period expenses of the operation. When combined with the cost of direct labor and standard material at each operation and work station, these standards will give a precise cost for each step in the production process. Accumulation of these costs into the inventory valuation will provide for precision costing of the product at every level of production and in its finished form.

H. GETTING OPERATING BURDEN INTO THE PRODUCT BUDGET

As discussed in the previous sections, the burden and overhead expenses will be accumulated and a burden and/or overhead "use rate" will be developed for each work station. This "use rate" will be stored in the cost system's data base or on a ledger card in a manual system.

As each part and operation is completed, the piece count and actual time will be recorded and entered into the computation system using either a computer or a manual procedure. The actual time will be multiplied by the standard "use rate" and the standard direct labor rate to produce the actual, totally absorbed, cost of the operation. The "use rate" and labor rate will also be multiplied by the standard operation time which is stored in the data base to compute the standard cost for the operation.

These two costs will be compared to compute any variances. The standard cost and actual costs will be posted into the inventory account to value the operation and cumulatively (with material cost added) value the product inventory to that point in the process. This same computation will be used at each step in the process to calculate the level-by-level work-in-process inventory valuation and the finished product cost.

The product cost budget and inventory valuation will be based on the standard cost calculations. The actual costs will be computed by multiplying the standard "use rates" and labor rates by the actual times and then adding the actual materials costs. The variances will be entered into a variance accumu-

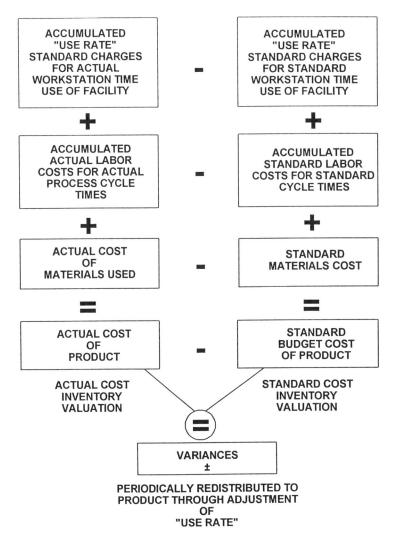

Figure 11.1 Product budget versus actual cost and the development of variances for periodical redistribution into the "use rate."

lation account and will be periodically fed back into the product cost through the space "use rate." The actual costs and variances will also be compared to the standards for each operation and element of cost to identify their causes and to make suitable adjustments to the processes or the standards. These comparisons (Figure 11.1) and variances will be reported to management on a part, product, operation, work center, employee, or department basis.

12

Facility/Building Burden or Overhead

A. FACILITY CAPITAL INVESTMENT VERSUS BOOK VALUE

In the development of the total absorption standard costing and the time use of facility based cost accounting philosophy, one must have a method for charging the product cost for the use of the enterprise's capital assets. This requires consideration of the accounting practices used for "writing off" or amortizing capital investments in an enterprise.

One of the conventional accounting practices which causes errors and distortions in the apparent cost of manufacturing operations is the application of depreciation expense to product cost. As discussed above, the IRS's permissible depreciation life and the allowable rapid write-off depreciation methods do not usually represent the true value, life, or cost of an asset. This is particularly true in the case of real estate investments and it also applies to many items of production and materials handling equipment. As a result, the book value of an asset in the financial accounting system may not have any relationship to either the current market value or the replacement value of the real estate or equipment item. This is an acceptable situation in the financial management world. However, it can result in a distortion of manufacturing and warehousing operating costs.

The book value which is used in financial accounting is computed as the balance of the initial capital value of the asset minus the depreciation accumulated from the purchase date to the present. In addition, the computation often accelerates the reduction in book value in order to gain tax advantages. From the manufacturing cost analysis point of view, this is an unrealistic method for valuing the facilities and erroneously charges variable asset based burden costs into the product cost. The facility and equipment expense should be uniformly charged into the product cost on the basis of its function or its contribution to the manufacturing or warehousing of the goods, and not on its book value. These charges can be applied to the product cost through the "use rate" which functions in the manner of a "rental" charge for the use of the capital assets.

B. LAND INVESTMENT AND NON-DEPRECIABLE ASSETS

From a financial accounting point of view, the value of land which is undeveloped or vacant is fixed and not depreciable. If it is sold at a different price from its original purchase, the price difference is treated as a capital gain or loss. However, the improvements on the land, such as sewers, utilities, and roads, are depreciable assets and should be included in the total valuation of the plant's building facilities. These improvements can sometimes be separated and depreciate at a different rate than that used for the buildings.

Another non-depreciable asset category is the enterprise's capital reserve. This may be in the form of investments or cash. These assets may be a consideration when computing the general and administrative expense's share of the facility "use rate." However, in this book, because their value fluctuates, they will be omitted from the definition of capital assets. The interest earned from these investments will also have an impact on the enterprise's profit and loss statement. These assets will not otherwise affect the product cost.

C. FACILITY/HOUSING EXPENSE, DEPRECIATION OR RENT

It is this author's belief that the logical basis for allocating the capital asset burden charge for buildings and equipment into the product cost would be to use a time use of facility "rental" fee or "use rate" based upon the approximate market rental value of the asset and the cost of its operation (Figure 12.1) This type of cost allocation can best be developed on the basis of a straight line depreciation charge coupled with the interest rate and operating expenses to produce a reasonable amortization of the asset and a continuing "use rate" or "rental" charge.

Among the expenses to be included in the facility/housing "use rate" would be the cost of maintenance and sanitation, utilities expense, insurance, taxes, security and fire protection, heating, lighting and air conditioning (HVAC), and real estate management. In cases where a part of the facility is leased from or rented to others, the earnings and/or costs from these arrangements should be treated as adjustments to the "use rate."

In the development of a total absorption standard cost approach to cost accounting, the author has established some arbitrary ground rules concerning

Facility/Building Burden or Overhead

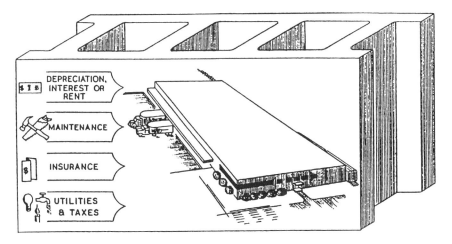

Figure 12.1 The building block of facility "use rate."

the handling of capital investment and its cost allocation to the product. These ground rules are summarized as follows:

1. Buildings and land, and their supporting utilities and improvements, will be assumed to depreciate on a straight line basis with a service life of 25 years (or the current IRS allowable period). In most cases, utilities investment will be included in the value of the buildings and be depreciated at the same rate. The interest charge on the purchase of land will be included in the annual cost of the building built on the land.
2. Manufacturing and warehousing machinery, equipment, and fixed tooling will be assumed to depreciate on a straight line basis with a service life of 10 years, or the allowable IRS service life for that equipment, whichever is least.
3. Interest rates to be applied to depreciation schedules and ownership cost will be set at the interest rate prevailing at the time of the purchase of the real estate or equipment asset.
4. Maintenance costs charged into the "use rate" for the real estate or equipment assets will be computed on the basis of the accumulated annual cost of actual maintenance divided by the scheduled annual working hours of the asset or, in the case of a new facility, a vendor estimate of the percentage of the purchase cost which should be allocated for adequate maintenance. Initial vendor recommended spare parts stocks should be treated as a part of the purchase cost of the equipment.

5. The sum of the annual depreciation charge, interest cost, utilities charges, and maintenance expense will be divided by the scheduled hours of operation of the plant, facility, or equipment as appropriate to develop the hourly "use rate" for the asset.
6. The capital asset "use rate" will be treated as a "rental" charge and will continue to be charged to the product cost as long as the asset is in use. The charge will not stop at the end of the allowable IRS "service life" of the asset! The charge will stop if and when the asset is disposed of.

Application of these ground rules will permit us to treat the "use rate" for buildings, facilities, and equipment in the same manner as a rental charge. This will have the advantage of providing a stable rate for accumulation of period costs into the product cost on a time use of facility basis. The "rent" for unused facility and equipment time will be charged to the enterprise's profit and loss statement and not to the product. The overabsorbed overhead and burden variance which results from operations in excess of the schedule will be treated as an addition to profit. The product will only be charged with the "rent" or "use rate" for the time during which it actually uses the facilities and equipment.

D. UTILITIES, MAINTENANCE, SANITATION, AND SECURITY

The cost of utilities and maintenance can be divided into two separate categories. Some of the utilities are period type costs which support the building and the working environment through lighting, heating, and air conditioning (HVAC), and provide sewage, potable water, fire defense water, and communications. Regular preventive building and grounds maintenance should also be treated as a period cost of the facilities. The expense of these services is a normal part of the building expense and can be included in the building's space "use rate" or "rental" charges. They are not linearly related to the production operation and those utilities which are production related should be treated separately.

Production related utilities include process water, process fuel (gas) and electricity, special fire protection and security, and production equipment maintenance. The sanitation (janitor) and security costs of the facility are partially related to the production operation. The need for security will increase with the hours of operation and the size of the work force. The need for sanitation (janitor services) will increase with the level of activity. These expense items should be treated as factory burden and not as a part of the building expense.

E. FACILITY UTILIZATION COST STANDARDS—"USE RATES"

In the development of facility cost standards, it is essential to first separate them into building or housing facility costs and production equipment or production facility costs. Although these two cost elements are later joined in the work station's "use rate," it is essential to separate them for identification of their contents. For the purposes of this section, let us designate "facilities" as including those items pertaining to the building and grounds, and not including the production equipment. Let us then designate the production facilities as "equipment" for the purposes of this discussion.

On the basis of the foregoing discussion, the annual cost of facilities or buildings would be:

$$B_A = D\%(B_{pc}) + i\%(B_{pc}) + i\%(A_{re}) + T_{re} + I_c + m_B + U_B$$

where:

B_A = the annual cost of owning and operating the buildings
$D\%$ = the annual straight line depreciation percentage rate for the IRS allowable life of the buildings
B_{pc} = the initial purchase price or cost of the buildings
A_{re} = the initial purchase price of acreage (land)
$i\%$ = the annual interest rate, at the time of purchase, on the money spent to purchase the buildings and land
T_{re} = the predicted annual real estate property tax rate on the predicted book or appraised value of the buildings and land in the middle year of the expected building service life
I_c = the annual casualty insurance cost (liability, fire, theft, etc.) on the building
m_B = the annual expected building maintenance cost.
U_B = the predicted annual fuel, electricity, water, and other utilities to support the building

The resulting annual building ownership and operating cost (B_A) is a part of the space based building cost standard or "use rate" calculation. By dividing this annual cost by the "net operating space" or the actual space used for manufacturing and warehousing, one can develop an annual cost per square foot of operating space. If this figure is then divided by the scheduled hours of operation of the plant, we develop a "use rate" per square foot, per hour of operation. This is the basic element of the overall "use rate" calculation.

As an example in the application of this procedure, let us use a 100,000 square foot building costing $50 per square foot and a five acre site costing $20,000 per acre. We will use a building life of 25 years, an interest rate of

10%, an annual property tax rate of 5%, an insurance cost of 5%, a maintenance cost of 5%, and a utility expense of 5%. Therefore, the annual cost of owning and operating the building will be:

$$
\begin{array}{lr}
100{,}000 \text{ sq ft} \times \$\,50 = \$5{,}000{,}000 & \\
\$5{,}000{,}000/25 = \$200{,}000/\text{annum} & \$\ 200{,}000 \\
\$5{,}000{,}000 \times 10\% = \$500{,}000/\text{annum} & 500{,}000 \\
5 \text{ acres @ } \$20{,}000 \times 10\% = \$2{,}000/\text{annum} & 2{,}000 \\
\$5{,}020{,}000 \times (4 \times 5\%) = \$1{,}004{,}000 & 1{,}004{,}000 \\
\hline
\text{Annual cost of the building} = & \$\ 1{,}706{,}000
\end{array}
$$

Applying the concept of the "net operating space," we would then divide this figure by 80% of the building area, or a similar factor. The resulting annual cost per square foot would be:

$1{,}706{,}000/80{,}000$ sq ft $= \$21.325$ per sq ft, per annum
$\$21.325/2080$ hours $= \$0.010252$/sq ft/hr $=$ space "use rate"
$\$0.010252$/sq ft/hr/60 $= \$0.000171$/min $=$ space "use rate"

or, using net operating hours $= 1952$ hours/annum,

$\$1{,}706{,}000/80{,}000$ sq ft $= \$21.325$/sq ft, per annum
$\$21.325/1952$ hours $= \$0.010925$/sq ft/hr $=$ space "use rate"
$\$0.010925$/sq ft/hr/60 $= \$0.000182$/min $=$ space "use rate"

A similar calculation is required for the development of the equipment "use rate." This procedure was discussed in Chapter 10.

Thus, two of the annual cost factors which must be included in the burden or overhead calculation are the annual equipment ownership and operating cost (E_A) and the annual building ownership and operating cost (B_A). The building ownership and operating cost is the basis for calculation of the space cost for each work station and the operation as a whole. These costs will become elements of the work station "use rate" which will be developed later.

13

General and Administrative Expense

A. DEFINITION OF GENERAL AND ADMINISTRATIVE EXPENSE

The general and administrative (G & A) expense account is often treated as the "stepchild" of the product costing structure. It is usually omitted from the "above the line" product cost or "factory cost of sales" and included in the "below the line" expenses. These below the line expense items are then buried in the profit and loss statement and do not show up in the product cost computation.

When G & A is included in the product cost, the cost allocation methods used to charge it into the product's manufacturing expense are among the most controversial in generally accepted accounting practices. A part of this problem lies in the definition of the content and role of general and administrative expense.

For the purposes of this book, the author offers the following concept of the role and content of G & A in a manufacturing or warehousing enterprise. We will use this concept or definition as the basis for including the G & A expense in the "use rate" computation and in the valuation of work-in-process inventory and finished goods.

Referring back to the philosophy of "total absorption" costing and time use of facility allocation of costs to the product, it is necessary to find a valid basis for assigning the G & A expense to the operation of a work station. This raises the need to define the true role of management in a manufacturing or warehousing enterprise. In the development of this engineered costing philosophy, we chose to use the following definition.

The responsibility and role of the "general management" of an enterprise is to *manage the use of the capital investment and assets* of the firm in a manner which will achieve the corporate business objectives and benefit the owners and shareholders.

General management fulfills these responsibilities by the effective administration of the utilization of the enterprise's assets.

It is therefore logical that the cost of management (G & A) should be distributed to the product through the product's time use of the assets. In order to accomplish this cost distribution, the cost of G & A must be related to the actual capital invested in the enterprise and the product's utilization of the plant facilities which represent this investment.

In most accounting systems, the G & A expense includes the fees and salaries of the board of directors, the president, and the other senior executives of the firm, the cost of their supporting staffs, and the expenses of operating the top management offices (Figure 13.1).

General and administrative expense usually includes the costs of legal and accounting services and staffs, computer support, outside consulting services, public relations, and memberships in trade associations. G & A will also include the cost of general liability insurance and the fees for financial management and accounting services.

Interest earnings from capital reserve investments might also be treated as a G & A income or a management cost reduction in some cases. Each company will define its own general and administrative expense account content.

Figure 13.1 The general and administrative cost block.

G & A Expense

If presented in a mathematical format, the G & A expense might appear to be:

$$G\&AX = X_{sf} + x_{sf} + Ox_{xs} + E_{xo} + B_{xo} + S_{pc}$$

where:

$G\&AX$ = general and administrative annual expense
X_{sf} = executive salaries, fringes, and bonuses
x_{sf} = executive office staff salaries and fringes
O_{xs} = executive office supplies and miscellaneous expenses
E_{xo} = executive office equipment ownership and operating expense (calculated with the same cost elements as manufacturing equipment)
B_{xo} = office building and land ownership and operating expense or rent (calculated with the same cost elements as manufacturing space)
S_{pc} = professional counsel (lawyers, CPAs, consultants, financial advisors, etc.)

All of the items in this G & A expense formula are on an annual basis. The G & A expense will be converted to an hourly basis for the computation of the "use rate." No allowance has been made for the interest earnings from liquid capital reserves. The reason for this omission will be discussed later.

In the case of a multi-plant or multi-division company, the allocation of an appropriate share of the corporate headquarters' general and administrative expense to the operating division or the plant level presents some complications. In many such situations, the G & A or overhead generated by the head office is charged to the local level in the form of a prorated annual or monthly fee. The fee share is usually computed on the basis of sales volume. In that format the charges are computed from a variable base. It would be better if the corporate headquarters' charges were computed on the basis of the capital invested in each division.

B. ABOVE OR BELOW THE LINE EXPENSES

In most generally accepted accounting practice, the general and administrative expense is treated as a "below the line" item on the profit and loss statement and in the product budget. As a result, the factory cost of sales does not include this expense. It is usually included in the markup or margin for pricing. The result is an erroneous product cost which does not include its fair share of *all* of the operating expenses of the company.

In the precision "total absorption standard cost" approach, we want to include all related costs in the computation of the product cost. It is desirable to include the general and administrative expense in the inventory valuation and product cost even though it is sometimes a very small element of the overall cost of operations. This approach raises the questions of how, why, and when to introduce the G & A expense into the product cost.

C. RATIONALE FOR INCLUDING G & A IN THE INVENTORY VALUATION

It is important to include all of the costs of the manufacturing enterprise in the computation of the cost of producing and distributing a product. This is necessary in order to be sure that all expenses are recovered by the profitable sale of the product.

In many accounting systems, the basic product cost calculation is accomplished by simply dividing all of the costs (or the burden and overhead costs) by the number of units produced. The result is a variable cost estimate which increases the apparent product value when production drops and decreases the value with rising volume. This is not a very realistic method of assigning overhead expense to the product, and it produces an invalid and inverse cost valuation.

It is better to use a time based, total absorption, engineered costing technique which develops realistic and precise cost estimates. This costing method is designed to accumulate all costs into the product on a time based, operation-by-operation (level-by-level) basis. The system should capture all of the related costs, but should not accumulate any costs that are not legitimately a part of the cost of making the product.

General and administrative expenses should be included in the production cost allocation procedure. These costs should be applied to the inventory valuation at each step in the process. This will give management the ability to measure the difference between a product's true manufacturing cost and its competitive market price. Knowledge of the true total cost of a product is essential in the making of marketing, manufacturing, and design engineering decisions.

As industry tends to become more capital intensive, the impact of labor cost on product cost decreases and the relative effect of overhead expenses increases. This includes the ratio of general and administrative expenses to shop floor labor costs. As stated earlier, the old one-third, one-third, one-third ratio of labor, materials, and overhead is shifting into a new pattern with as little as 5–10% for manufacturing labor and increasingly higher cost shares for materials and overhead.

At the same time, industry is investing in expensive capital equipment to improve quality and performance. This results in a rising depreciation expense allocation, an increase in the cost of owning capital facilities, and an expanding technical and executive support role at the highest management levels.

Management must oversee the use of all of these investments and expenses. It is essential to capture all of these management costs, as they must be included in the calculation of the true product cost.

An apparently obvious approach to achieving this relationship with a stable base would be to distribute the G & A expense on the basis of the net operating space and the scheduled production hours used in the operation. However, this technique is inherently erroneous because the uniform distribution of space cost does not recognize the non-uniform distribution of capital investment in assets such as production machinery and materials handling and storage equipment.

In addition, the space approach can only be equitably applied in a single plant enterprise when the executive offices are a part of that plant. In a multiple plant operation, the variations in each plant's equipment investment would not be recognized and the distribution would be erroneous.

For these reasons, and in keeping with the belief that it is management's job to manage the capital investment on behalf of the owners, it appears to be desirable to relate G & A to the invested capital in use within the manufacturing and warehousing operation.

The definition of invested capital is a critical element of the system. In designing the precision costing procedure, *invested capital* has been defined as including buildings and land, manufacturing and materials handling machinery and equipment, office equipment and computers, and engineering and laboratory equipment. Capital invested in liquid assets such as inventory and securities has been omitted because its value varies with time and it therefore contradicts the basic cost system objective of achieving a stable cost base. The use of fixed assets as the basis for defining invested capital assures a stable base for development of a standard general and administrative expense "use rate."

In a pure sense, and in mathematical form, definition of the invested capital would be:

$$C\$ = B\$ + E\$ + S\$ + c\$ + WIP\$ + FG\$$$

where:

$C\$$ = total invested capital
$B\$$ = purchase in-place cost of buildings
$E\$$ = purchase in-place cost of equipment
$S\$$ = Purchase cost of securities investments.
$c\$$ = Cash on hand or working capital.
$WIP\$$ = Dollar value of current work-in-process.
$FG\$$ = Dollar value of current finished goods inventory.

This *total value* of the invested capital (not the book value) can then be used as the basis for distributing the G & A expense into the work station "use rate." However, as stated above, the liquid assets and inventory values vary with time and can distort the validity of a standard cost "use rate." Therefore, the definition of invested capital which is to be used in the precision costing system will omit these variable assets. It would be:

$$C\$ = B\$ + E\$$$

The G & A will then be treated as an "above the line" expense in the costing of the product.

The G & A element of the hourly work station "use rate" (U_{ga}) is defined as:

$$\frac{[G\&AX\ (\$/annum)] / [C\$\ (\$\ \text{in-place capital cost})]}{SOH(\text{scheduled operating hours per annum})}$$

where:

U_{ga} = the G & A portion of the plant's hourly "use rate"
$G\&AX$ = annual G & A expenses
$C\$$ = total landed and/or in-place investment cost of the fixed capital assets of the plant (not book value).
SOH = scheduled operating hours per year

D. ALLOCATION OF GENERAL AND ADMINISTRATIVE EXPENSE

Following the logic in the foregoing paragraphs, the G & A expense can be allocated to each work station in proportion to the value of the capital invested (C$) in the equipment, the space utilized by that work station, and the time use of that facility investment (t).

This will require allocation of G & A to the work station on the basis of the dollars of G & A expense per dollar of capital invested in the work station's space and equipment. This figure, divided by the scheduled hours of production (SOH), produces the G & A "use rate" cost per hour, per dollar of invested capital.

This hourly G & A "use rate" figure is then multiplied by the total dollar value of the fixed investment in equipment and space for the work station to compute the hourly G & A "use rate" (U_{ga}) for the work station.

In mathematical format, the hourly G & A "use rate" would be:

$$\frac{G\&AX/C\$}{\text{Standard hours of operation}} \times C\$_{ws} = U_{ga}$$

where:

> G&AX = $'s of annual G & A expense
> C$ = $ of total invested capital @ original cost of fixed assets
> C$$_{ws}$ = capital invested in work station's space and equipment
> U$_{ga}$ = G & A share of hourly "use rate" for work station

E. G & A COST APPLICATION STANDARDS

This calculation may result in a very small "use rate" figure. In some cases it may be less than $0.00001 per dollar of invested capital per hour. This suggests that G & A may sometimes have a negligible effect on the computed cost of the product. However, in capital intensive operations, computation and application of the standard G & A "use rate" may be critical in determining the correct cost of the product.

For example, if a $200,000 machine is located in a work station which occupies three hundred square feet of $50 per square foot space, the total capital investment in the work station would be $215,000. Using the above sample G & A standard of $0.00001 per dollar per hour, the G & A charge to the work station's "use rate" would be $2.15 per hour or approximately $0.036 per minute. This is not a negligible cost item!

This G & A cost standard must be included in the work station's standard "use rate" in order to compute a "total absorption standard cost" for each operation and for the products produced by the operation. As the product progresses through the process, the G & A expense will become a part of the cumulative inventory valuation which is developed at each level or step in the manufacturing system.

F. GETTING THE G & A INTO THE PRODUCT BUDGET

The product budget is the total of the accumulated costs of labor, materials, burden, overhead, G & A, and facility expenses which are generated during the manufacturing and distribution process. The burden, overhead, facility expenses, and G & A expense are accumulated into a work station "use rate" standard cost. This standard is charged to the product for each hour or minute for each operation which is performed in the work station. The product budget is a level-by-level statement of the accumulated costs generated by each step or operation.

In the total absorption standard cost, time use of facility, manufacturing cost accumulation approach, expenses are entered into the value of the work-in-process inventory on a level-by-level, time use of facility basis for each operation. This is accomplished by reporting the standard material cost, batch quantity, process time, labor rate, and "use rate" charges into the computer or manual inventory record for each operation at each step in the process. The system is designed to take these data and compute the accumulated work-in-process or finished product value at each level. The computation will produce both the standard and actual cost of each operation and any variances.

A key tool for accomplishing this cost accumulation is the work station's "use rate." This is the "bucket" into which all of the indirect and overhead costs are gathered. The G & A "use rate" (U_{ga}) is included in the work station's overall "use rate" which is charged to the part or product on a time use of facility basis.

In effect, this is an "Activity Based Costing" (ABC) technique! It charges *all* costs into the product on the basis of the time consumed by an "activity" or operation, at each work station, and for each production setup.

14

Precision Costing of Manufacturing Operations

A. THE CONSTRUCTION OF THE "USE RATE"

The work station's "use rate" is a time based cost unit which includes its time related share of all of the space, equipment, indirect, overhead, period, general and administrative, and miscellaneous costs of the manufacturing enterprise. As stated earlier, the only incorruptable elements of the cost equation are time and space. The "use rate" is designed to accumulate the appropriate time based segment of all enterprise costs into each square foot of the net operating space used by the manufacturing operation or activity.

Reviewing prior comments, the "net operating space" is defined as the area utilized by the manufacturing operation and its supporting aisles (or the warehouse, storage space and aisles). It does not include the management offices, toilets, entries, stairs, lobbies, cafeterias, lockers, and other supporting service areas. The costs of owning and operating these areas are absorbed into the net operating space's "use rate."

All of the annualized indirect operating expenses, period costs, utilities charges, maintenance expenses, indirect materials, manufacturing overheads, taxes, insurance, and facility (building) ownership costs (depreciation or rent) are totaled and divided by the "net operating area" to compute a space "use rate" per square foot, per year.

The "use rates" for capital equipment which is building oriented, such as heating and air conditioning machinery, compressors, cranes, generators, fire pumps, transformers, security systems, etc. are added to the net operating area's space "use rate" on an overall cost per net square foot, per annum basis. This produces the net operating area's total annual space "use rate."

This figure is divided by the scheduled annual hours of work (not including holidays and vacation shut downs) to produce the hourly "use rate" per square foot of net operating space. This is the basic vehicle or "bucket" into which other "use rate" charges are added to produce the complete space "use rate."

The space "use rate" for the work station is computed by multiplying the area of the work station (including its safety, maintenance, and working space) plus one half of the aisle which services the work station (or all of the aisle if the opposite side of the aisle is a wall), by the net operating area's total hourly space "use rate" per square foot.

In a separate calculation, the depreciation or lease charges, and the annual insurance, interest, power, maintenance, and taxes, on each unit of manufacturing machinery, or production capital equipment, are summarized and divided by the scheduled operating hours for the individual production machinery unit to compute the annual and hourly "use rate" for that unit of machinery or equipment.

The individual "use rates" for manufacturing machinery, special tooling, fixtures, inspection equipment, work station specific materials handling equipment, and other process related equipment in each work station are added to the work station's space "use rate" for the work station in which the equipment is located.

In addition to the combined hourly space, equipment, and machinery "use rates" for each work station, the "total absorption" costing system also allocates a fair portion of the general and administrative expenses (G & A) to the product through the work station and its operations. The method for calculating this G & A "use rate" was discussed in Chapter 13. Its inclusion in the work station's "use rate" is accomplished by multiplying the hourly G & A "use rate" (in terms of G & A dollars, per invested dollar, per hour) by the total capital invested in the work station for space, machinery, special tooling, and operational furniture.

The value of the capital invested in a work station is computed by adding the dollar amount resulting from multiplying the initial, per square foot purchase cost of the building and its appurtenances by the area of the work station to the initial price paid for all the production machinery and equipment used in that work station.

Therefore, the total hourly "use rate" for each work station includes the time use costs of the work station space, manufacturing equipment, overhead, indirect and period expenses, and its share of the G & A expenses. This is shown in a graphic model form in Figure 14.1.

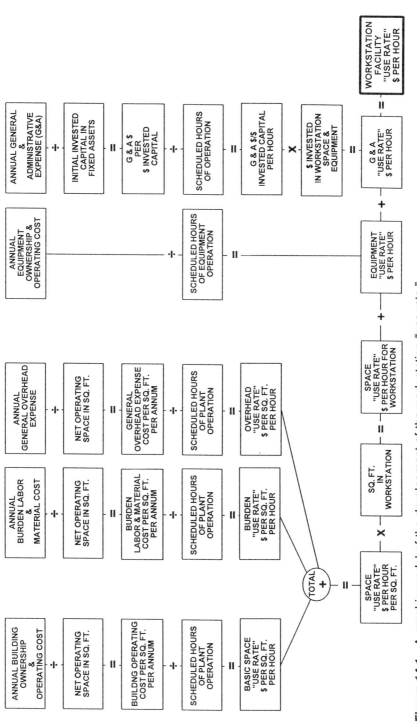

Figure 14.1 A graphic model of the development of the work station "use rate."

B. APPLYING THE "USE RATE" IN INVENTORY ACCOUNTING

As stated several times in the foregoing chapters, the basic concept or philosophy of the time use of facility and "use rate" approach to indirect and period cost absorption accounting is the same used when computing the rental charges for real estate, canoes, automobiles, or construction equipment. In each case, the time based rental charge includes all of the hired asset's fully absorbed ownership, management, and maintenance costs. In most cases, the commercial rental rate also includes a profit for the owner. However, profit is not included in the computation of the proposed cost accounting "use rate."

This same product "rental" concept has been applied in the development of the "time use of facility" and "total absorption standard cost" approach to product costing. By building an hourly "use rate" charge for each square foot of the facility, each production machine, and each work station, we can provide a vehicle for the capture, accumulation, and assignment of applicable indirect, overhead, and period costs to each manufacturing, materials handling, and processing operation for each individual product. This "use rate" charge is accrued on the basis of the "time use of the facility" (work station and machinery) and the time used by each process operation. These accumulated costs are added into the cumulative value of the work-in-process inventory after each step in the manufacturing process.

This procedure permits cumulative total absorption of the correct share of the indirect, overhead, and period costs of the enterprise into the work-in-process and product cost on a level-by-level, time use of facility basis. When the cumulative "use rate" charges are combined with the standard or actual material cost, and the accumulated labor charges after each operation, the resulting fully absorbed work-in-process inventory value will be accurate at every step in the manufacturing process. This will result in an accurate, activity based, finished product cost.

C. PRODUCT COSTING IS INDEPENDENT OF PRODUCTION VOLUME

When the "use rate" is applied in the computation of manufacturing costs, the operation, part, or product is only charged for the "use rate" (the work station "rent") during the actual time or "activity" when the work station and its machinery are used to perform the actual manufacturing operations required to make the part or product. This operation or "activity" is the only basis for accumulating costs into the work-in-process inventory or product. In effect, this is a precise form of "activity based costing" (ABC). It is not dependent upon, or related to, production volume.

The work station's "use rate" captures its time and space based share of all of the facility and management related expenses of the whole manufacturing enterprise. The proper share of these costs is allocated to each part or product in proportion to the time used in the performance of each manufacturing operation or activity, in each work station, during the production time used for each individual part or product. When the work station is idle, its "use rate" charges are allocated to the profit and loss accounts of the enterprise. The part or product is not charged with the cost of idle work station time. Overabsorbed burden and overhead which results from overtime or longer work schedules will show up as a reduction in the burden and overhead accounts or as a profit for the enterprise.

As a result, *the costs allocated to the part or product are independent of the volume or activity level of the plant as a whole.* Only the material, labor time, and the work station "use rate" time used in the actual operations or activities required to produce the parts or product are charged to that part or product. Each part or product produced in the manufacturing system bears its share of the whole cost of the enterprise, but no more than its share. *Variances in production volume and capacity utilization will have no impact on the unit cost of the part or product.*

D. PRECISION COSTING BY PART, OPERATION, DEPARTMENT, EMPLOYEE, AND PRODUCT LEVEL OF COMPLETION

As pointed out in Chapter 7, a numerically indented manufacturing bill of materials and a matrix type part number coding system are very important components of the total absorption, level-by-level, precision costing system. It is possible to write a computer program to perform the same functions, but the logic of the matrix coding system and numerical bill of materials must be built into the program. In the total absorption, precision cost system, the matrix coding structure and the numerical bill of materials must perform the following functions.

1. The bill of materials and the coding structure will jointly provide matrix type random access to individual parts' data at each level of completion
2. The coding system will precisely and individually classify, identify, and define, each part, semi-finished part, sub-assembly, assembly, and product used in the design and manufacturing system.
3. The coding system will be capable of level-by-level inventory access to all parts at each level of completion.
4. The system will permit the vertical accumulation and listing of all parts and components used in a product in order to compile a bill of materials.
5. The numerical bill of materials structure will allow access by part number to produce "where used" listings of the manufacturer's finished products which use common or standard parts.

These access patterns are required in order to retrieve the identity, completion status, and cost of any part, component, or product from the inventory system. This retrieval capability is critical to the development of product or part cost estimates and in the assignment and accumulation of manufacturing and storage costs into the inventory valuation of a part, component, or product. As stated above, advanced computer spread sheets and custom software can be used to accomplish these tasks without using the SIMSCODER numbering system if the same type of matrix access logic is included in their algorithms.

Another key element of a precision costing system is a well defined operation sheet or routing which describes the steps or operations in the manufacture of each part and the sequence of the assembly of the product and its sub-assemblies. A well constructed operation sheet usually includes operation numbers, a preferred work station or machine type, an identity number for each operation, and a standard or estimated operation time for each setup and operation. The operation sheet may also include tooling data, material requirements or standards, and process instructions.

In most cases, each operation represents a level of completion in the manufacturing process and a level of work-in-process control in the materials management system (Figure 14.2)

The coding structure presented in Chapter 7 fulfills the information identification and control requirements of the operating system. It is called the "Structured Information Management System for Coded Operating Data and Engineering Retrieval" or "SIMSCODER." This matrix coding system is similar in function to a telephone number (Figure 7.14). It shrinks in size as it focuses on a target item and it can be accessed at several levels. (Figure 14.3)

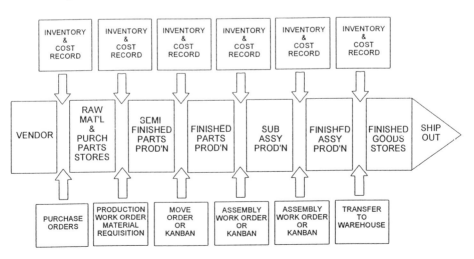

Figure 14.2 The process flow/material management relationship.

Precision Costing of Manufacturing Operations

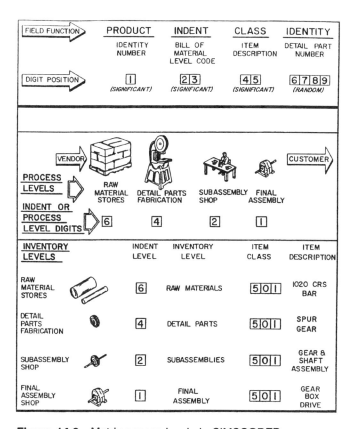

Figure 14.3 Matrix access levels in SIMSCODER.

This basic code concept and structure can also be applied to employee clock numbers, machine and work station identity numbers, department numbers, and operation numbers (Figure 7.17) as well as many other items of information as shown in Figure 14.4.

In addition to the proposed SIMSCODER part numbering and bill of materials structures, there are also some commercially available "bill of materials processors" and materials management software packages which can accomplish all or part of these information management functions. However, the problem with many of these is that they do not easily accommodate the level-by-level access and cost accumulation system recommended for the proposed precision costing program.

By referring to the coding system discussion presented in Chapter 7, the following commentary on precision costing will assume that the level-by-level cost system has made practical use of a similarly structured matrix coding sys-

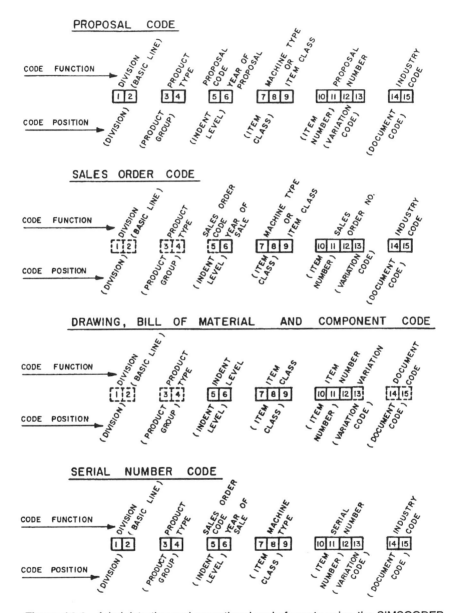

Figure 14.4 Administrative and operational code formats using the SIMSCODER technique.

tem and a level-by-level numerically indented manufacturing bill of materials structure.

The key element in the definition or estimation of the cost of a product, or in the computation of a level-by-level, work-in-process inventory valuation, is the completed individual part from which the product is assembled. The part cost is composed of the costs of materials, labor, and the time use of facility's "use rate" charges which represent the manufacturing burden and overhead. The accumulated part costs are applied to the WIP inventory valuation and to the costed bill of materials for use in computing the manufacturing cost of the product.

In addition to the cost collection for product costing purposes, management also has the need for detailed operating and cost data to measure the performance of the manufacturing process. It is therefore essential to precisely define the actual variances from the predicted utilization standards for cycle time, materials, labor, and facilities' "use rate" for each operation, and for each part being manufactured. Identification and capture of these variances provides the source data for continuous improvement of the process. In order to accomplish this, the operation number must be tied to the part number at each step in the process.

It is also desirable to identify the relationships of these variances to the lot or order number, employee, work station, and the department involved in the operation. Figure 7.12 shows this data collection and analysis concept.

The operation sheet or routing, or its computer equivalent, usually includes the part, operation, and work station identity numbers in its data content. By combining or matching these data with the employee identity number and order number, one can extract the cost of each operation for each part and each order or lot. These costs can also be assigned to the valuation of the WIP inventory. These cost data can also be compared to the assigned standard costs to define the variances caused by the choice of a different operator or the choice of an alternative work station or machine.

For example, if a part is manufactured on a machine with a higher "use rate" charge than the operation sheet calls for, the variance in the "use rate" will be caused by the shop schedule. Likewise, if a higher rated employee is running a job on a lower rated work station or machine, the employee will be paid at his usual rate and the variance in the job's labor cost will be caused by the staffing decision. In either case, the process time may be the same and the variances will be unrelated to production volume.

This is in addition to defining the variances caused by the actual versus standard work cycle time and/or the material's cost or quantity used. These costs can be computed at each operation or process step to provide for a level-by-level costing of WIP parts, assemblies, and finished products. A comparison with the predicted and listed standard cost will define the variances.

In summary, the use of coded data elements, and their collection via manual means or into a computer data base by keying, bar coding, or other means, can provide the basis for computing and defining the cost of each step or operation for each part, at each work station, and for each lot or order, each day (or hour).

This procedure will also allow management's measurement and evaluation of each employee's performance on each operation and each item. It will permit the comparative measurement of the performance of each machine or work station on each operation and on each part or item. In serialized operations, the procedure can be applied at an assembly line or manufacturing cell level. The assembly line as a whole, or the manufacturing cell, can be treated as a work center for costing and scheduling purposes.

Accumulation of the operations' costs by work station and by department will also provide the basis for responsibility accounting at the department and/or division level. These same work station data elements also provide the information input for activity based costing (ABC) and analysis. Figure 7.12 shows the type of management information that can be extracted.

These capabilities permit detailed costing and analysis of the operation, precise cost estimating of parts and products, and provide a basis for cost effective design and value analysis in product development.

E. GROUP TECHNOLOGY AND PROCESS SEQUENCE COSTS

Manufacturing people have long recognized that the proper grouping and sequencing of parts and products passing through a manufacturing process or work cell can reduce setup and job change time and the overall cost of production. This concept is the basis for multiple part setups on turret lathes and automated turning machines; multiple part drill jigs on multiple spindle drill presses; and multiple part fixtures for machining centers and milling machines.

For example, when turning different gear blanks from a single bar of stock, the sequencing of a bevel gear blank through the turret lathe before the spur gear can reduce both process change time and work cycle time. This is because the spur gear will use the same roughing and blanking operations and simply omit the bevel cuts with only dimensional alterations and no setup changes to make the spur gear blanks.

In today's state-of-the-art, computer numerically controlled (CNC) manufacturing environment, which includes machining cells, flexible manufacturing systems (FMS), tool changers, automated machine load/unload systems, and lot sizes of one or a few pieces, the importance of grouping parts and products which use similar processes has become more widely recognized. When similar parts and processes are sequentially scheduled through a work station, the programming of the CNC machines and the loading of the tool changers can be more efficient. In the process, the changeover from one part to another will require less alteration in the operation sequence or tool utilization. Multiple part fixtures can be pre-loaded off line to reduce idle machine time for part changes. In bar fed operations, the same material can be used for a series of different, but similar parts.

For many years, manufacturing engineers have been attempting to apply the concept of "group technology" to this situation. They have logically and visually grouped products requiring similar operations and batch fed them through machining cells and processes. As a by-product of this effort, many attempts have been made to develop coding systems which define the shapes and processes in a manner which will allow computer sorting of parts orders into process oriented groupings. Most of these coding systems have, in the opinion of the author, gone into too much process and material detail and have therefore been cumbersome and difficult to apply.

It is the author's belief that the geometry of the part can be coded in a manner which will allow the grouping of the items in accordance with their required manufacturing processes. The processes can then be sequenced by the manufacturing engineer or by an artificial intelligence (AI) algorithm. This concept has been successfully applied.

A further grouping can be achieved by also sorting the parts by their functional definition. As noted above, the SIMSCODER has a function field as part of its structure. This can be used to accomplish the functional sorting of the parts.

By combining geometric, functional, and size sortation, the objectives of group technology can be achieved in a relatively simple manner. The author's graduate student, Ricky Anderson, tested this combined geometric and functional coding concept in his 1993 Master's Thesis at Ohio University. It was found to be a simple and practical basis for applying the principles of group technology.

The purpose of group technology is to sequence parts through work stations, machines, or manufacturing cells in a manner which will reduce setup cost, minimize changeover time, and allow reduction of batch sizes; thereby reducing the WIP inventory. The result will be a lower unit production cost and a lower product cost.

In many cases, management must evaluate the tradeoff between more efficient manufacturing based on group technology, and the possible increase in work-in-process inventory caused by the early manufacture of parts which fit into the process group. The availability of a precision cost procedure will simplify this evaluation, improve its accuracy, and help in making better scheduling decisions.

F. PRECISE MATERIALS MANAGEMENT IN A JUST-IN-TIME ENVIRONMENT

A primary objective of Just-in-Time (JIT) manufacturing or logistics management is to minimize work-in-process inventories without creating work stoppages caused by material shortages. This can be accomplished by reducing the volume of material moved in each transaction, increasing the transaction rate, reducing the setup times between operations, and maintaining tight material control.

In theory, this is a simple matter of moving small quantities of material and/or work-in-process (WIP) from one work station to another each time it is needed in the next operation. By applying the KanBan card's "pull" system, or a computer based shop floor control program, one can manage these movements and normally guarantee the availability of materials ahead of each operation without excessive WIP inventory. This system works best in a serialized manufacturing system where all components can be scheduled together. It is more difficult to apply in a "job shop" or random variable product production environment.

In addition, when applying JIT, one must address such problems as imbalances between machine rates and production schedules, the reliability and "up time" of production equipment, the impact of group technology on WIP requirements, quality standards, supplier reliability, and the space required for large components.

Another cost factor is the impact of JIT on materials handling. Just-in-Time's reduced volume per movement, and increased frequency of movement, contradicts one of the basic principles of materials handling engineering.

Materials handling principles direct "movement of the largest possible quantity, as far as possible toward the next point of use before putting it down or breaking up the load." The JIT concept demands the opposite approach! In JIT, we move small modular quantities (usually one or two tote boxes or one pallet, and often only one part) from work station to work station, usually in short moves, and often by conveyor. Application of automatic guided vehicles, conveyors, and robots can eliminate labor cost from these material movements, but they still have a transaction cost.

In a JIT operation which uses modular or "common denominator" movements in tote boxes or on pallets, application of bar coding, radio frequency (RF), or other types of automatic identification can improve the precision of materials management and cost data collection. For the best results, the quantity in each handling module or "common denominator" unit should be standardized. Each handling "common denominator" should be tagged with a bar code, RF, or other automatic identification device. The product or part being moved should also be individually bar code marked, and the quantity per module should be standardized for each part.

With these identification and standardization practices in place, movements can be precisely monitored by automatic identification devices in either a mechanized or manual wanding system. This will provide real time flow status reporting for the materials management system. Cost data can also be accumulated in the WIP inventory record at each move. This will result in precise WIP control, level-by-level cost reporting, and flow control. The WIP inventory's quantity and valuation can be updated after every movement and production or value adding operation.

These same data acquisition methods and "common denominator" handling practices can be applied in job shops or non-JIT operations with equal effect. However, if the handling unit quantities are not standardized, counting will be required at each step in the flow. This can sometimes be accomplished by computer connected weigh counting stations, or by photoelectric cells or lasers on a conveyor system.

Although we strive for "zero defects," few operations are devoid of rejects or scrap. In a JIT situation, rejects can force the generation of "fail-safe" safety stocks ahead of sensitive work stations or bottleneck operations, cause extra material movements, and disrupt the modular quantity control system. Errors or rejects can also shut down the operation.

It is therefore necessary to capture, identify, and report all rejects and scrap, and to make in-flow quantity and schedule adjustments to assure compensation for any imbalance or reduction in stock availability. The rejects must also be "backflowed" for either rework, return to vendor, or scrap.

All of these transactions must be recorded into the inventory and properly costed. Scrap costs will be treated as a material cost variance and rework costs will be considered an operational variance when they are not predicted and included as a part of the standard material cost.

Precision material control can be simple in a JIT environment when modular "common denominator" material movement systems and automatic identification procedures are applied. These material management methods also work well in job shop or non-serialized environments.

G. PRODUCT DESIGN DECISIONS AND PREDICTABLE COST DATA

Modern manufacturing managements are attempting to apply computer supported "concurrent engineering" and "design-to-cost" techniques as a means of shortening the lead time from concept to market and as the basis for product design, marketing, and manufacturing policy decisions. If the cost and producibility of a product is definable before it is fully designed and tested, management can evaluate its marketability and pricing before it is manufactured. These techniques can also help to reduce or minimize prototyping costs. If the engineered precision cost system is the basis for the estimating data, the estimate will be independent of volume forecasting errors.

These engineering management concepts are not new! Concurrent engineering and design-to-cost practices were difficult in the past. However, concurrent product and process engineering was often attempted by having the product designer and manufacturing engineer work as a team. They often shared an office. Computer aided design (CAD) software and hardware, and computer based cost analysis and estimating systems have made concurrent engineering more feasible. Computer aided design, computer graphics, networking, and the ability to download and convert a part's design directly into the machining software for CNC equipment, have provided the tools for state-of-the-art concurrent product and process design practices.

The availability of detailed and precise, level-by-level, manufacturing operations information, and precise, level-by-level, part, component, and process cost data in a computer accessible data base supports concurrent engineering and design-to-cost operations. It also provides an almost "real time" data source for both product design decisions and manufacturing methods development. In fact, recognition of this need for detailed data retrieval during the 1960s was one of the motivating factors leading to the author's early development of the SIMSCODER part numbering system and the concept of time use of facility, total absorption standard costing.

These part coding and manufacturing costing techniques provide the language and information format structures for the retrieval of precise and detailed cost data for the computation of manufacturing costs on a part, item, and operation basis.

This capability makes the design to cost procedure more effective and reliable. By utilizing retrievable cost and operations data details from the computer data base, the concurrent engineering team can continuously test the part and process designs against alternative product designs and the product cost budgets or pricing targets. These detailed cost and process data permit comparison of different part and product designs and alternative manufacturing methods during the concurrent engineering process. This allows management and engineering decisions to be made on the basis of reliable cost estimates.

Reliable cost estimates can be constructed from actual data drawn from prior similar designs and operations, from a computer stored library of time standards and shop tested operational procedures, and from part and component values drawn from WIP inventory records. The manufacturing engineer can also use the computer to assemble and manipulate historical data, machine tool vendor standards, predetermined time standards, and operations data elements to develop estimates of the time and cost of a machining or assembly process.

The engineer can simulate and test the part design and the process by applying the CAD design to the machine control (CNC) program or by applying manual process planning and estimating techniques. The manufacturing simulation can then be cost estimated. The required design and methods decisions leading to initiation of manufacturing can be based upon the predictable product cost projections produced by the manufacturing process simulation.

H. MARKETING AND PRICING DECISIONS WITH PREDICTABLE COST DATA

When reliably predictable cost estimates are available to management, they can be used as the basis for making marketing and pricing decisions. However, the price is normally established by the market and the competition. It is therefore essential to determine the expected product cost with reliable accuracy before attempting to set a competitive price for entry into the market.

If the cost is too high to allow an acceptable margin in the face of competition, the product design and manufacturing process must be reevaluated to identify the possibilities for cost reduction without loss of quality, performance, or customer acceptance. In such a case, the design-to-cost process will apply many of the principles of value engineering to modify and/or redesign both the parts and the complete product. The principles of industrial and manufacturing engineering will also be applied to reduce the cost of the process.

When an optimum cost has been defined, management can make the marketing and pricing decisions concerning the wisdom and potential profitability of releasing the product into the market.

I. BETTER RESPONSIBILITY AND ACTIVITY ACCOUNTING

In a manufacturing environment, there are several purposes or reasons to operate a good cost accounting system. Some of these reasons have been implied in the preceding discussions of design engineering, manufacturing policy, and marketing decisions. The need for good manufacturing cost data in the development of pricing policy is also obvious. Another reason for using good accounting practices relates to the need to accurately define costs and profits for financial management and taxation. In addition, good cost data is essential to capital facilities investment decisions and operations management and control.

In recent years, manufacturing management has shown considerable interest in the financial philosophies of "responsibility accounting" and "activity based costing" (ABC). In each of these concepts, the objective is to relate costs, and their variations from plans and standards, to the portion of the organization which has responsibility for the activities which generate or control the costs.

In responsibility accounting, the focus is on the organizational source of the costs. In activity based costing, the focus is on the operation or function which generates the cost. In both cases, the costs originate in a work station or work center. In cases where generally accepted accounting practice is used, the captured costs include labor and materials but often relate burden and overhead to either labor or production volume and leave general and administrative expense below the line.

Conversely, in the proposed time use of facilities and total absorption standard cost approach, the work station's operations costs include all of the elements of burden, overhead, labor, and materials as we have discussed.

We have addressed the capture of the product's manufacturing costs on an operation-by-operation or level-by-level basis within each work station, and for each part or component. This approach provides for the creation of a detailed and cost defined data base in an operational format. Each item in the data base can be related to the operation, the work station, the part, the employee, the work order, the product, and the date. As a result, each item in the data base can also be related to the organizational entity which governs the operation, work station, or department, and to the activity or operation being performed.

In addition, as shown in Figure 14.5, the proposed time use of facility and total absorption standard cost approach is independent of production volume or capacity utilization. It relates the cost of the product or part to the operation and work station, thereby giving a clear basis for organizational responsibility accounting and/or activity based costing.

This approach also provides a measure of the share of the manufacturing capacity used in any period of time. This allows management to measure the cost of the unused capacity and to place the responsibility for the unused capacity cost on the "activity" which really caused it—the failure of the marketing and sales operation to capture enough business to fully utilize available capacity.

Manufacturing does not control the amount of work which it is given. That is the function of sales. Manufacturing only controls the efficiency with which it performs the work when assigned. This is measured by the variances from plan or standards, not by the share of available capacity which is used.

The proposed time use of facilities, total absorption standard cost system allows management to measure and monitor the complete manufacturing cost and the operation's performance effectiveness on a detailed product, part, operation, employee, machine, work station, department, supervisor, order, or lot basis. This technique provides the cost and data elements required for either responsibility accounting based on a department or executive's performance, or for activity accounting based on an employee, machine, work station, or process's performance.

Precision Costing of Manufacturing Operations

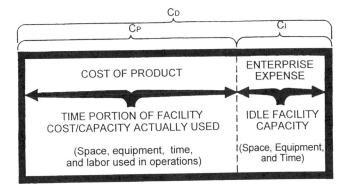

C_D = Total burden and overhead for whole facility

C_P = Total burden and overhead for actual time use portion of facility

C_I = Total burden and overhead for idle time portion of facillity

C_I should be charged as an unabsorbed variance to profit, sales or management expense, not to product or manufacturing expense

Figure 14.5 The basic concept of the time use of facility approach to total absorption standard costing.

This system also measures performance on a total cost basis for each product or activity, and for only that portion of the plant capacity which is in use. At the same time it provides a means for defining the cost of idle capacity and can place the responsibility for idleness on the cause—the failure to sell enough work.

15

Precision Costing of Warehousing and Materials Handling Operations

A. APPLYING THE "USE RATE" TO STORAGE SPACE

As previously stated, the "use rate" costing concept is designed to apply a rental philosophy as the basis for charging indirect and period expenses into the cost of manufacturing a product. This same approach can be applied to the cost of storage with some variations in the procedures and in the definition of the "use rate."

When one addresses the cost of storage and materials handling, it is obvious that there are no changes in the products being stored and handled and no value is added to the product. Storage and materials handling are all-cost activities.

However, the cost of materials handling and storage operations must be captured and accurately included in the computation of the "landed cost of sales" of the product. In dealing with this expense, the internal handling and storage costs within the factory should be treated as an addition to the manufacturing cost.

In some cases, the finished goods warehouse and the shipping operation are considered to be a part of the distribution cost. However, in this chapter, and for the purposes of explaining the system, the finished goods warehouse and shipping operation will be treated as parts of the total warehousing operation of the manufacturing facility. A possible free-standing distribution warehouse will be addressed separately.

If the material or product is safely packed and handled in a "common denominator" module or unit (tote box, pallet, carton, crate, barrel, bag, etc.), the identity of the material has only marginal impact on the storage and handling cost. With the exception of security, fragility, perishability, insurance, and ownership interest cost, the value of the product being stored has no effect upon the cost of storage or the materials handling operations. These "soft" or intangible product characteristics are of little concern in the design of the materials handling and storage operations. Storage costs are a function of cube, shape, and time. The value of the goods being handled has *no effect* on the storage or handling cost!

Nonetheless, the "dollar density" of the goods in storage has an effect on the economic efficiency of the inventory policy being followed by management. A reduction in the package cube (or an increase in its dollar density) will improve the economics of handling and storage operations by increasing dollar throughput and cash flow for greater profits (Figure 15.1).

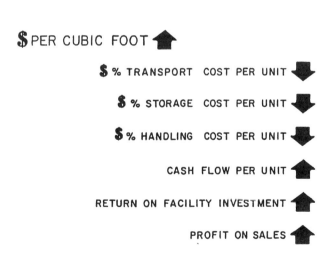

Figure 15.1 The dollar density effect.

Warehousing and Handling Operations

The critical issues or product characteristics in the design of a storage and handling system are the size or cube of the product, the dimensions, shape, weight, and stackability, of the product, and the number of units of product which will normally fit on or in a "common denominator" (tote, pallet, package, etc.). For the design of the handling system, consideration of the transaction rate or material movement pattern is also a critical factor.

From a philosophical point of view, and assuming that the storage and handling equipment has been selected to provide adequate capacity and is not weight limiting, the key cost capturing element in a storage or handling system is the cubic foot of product or its cubic feet of space occupancy. The pallet, tote, storage rack, warehouse, or storage machine handles cubic feet of product, not inventory items. The "common denominator" makes the product's identity "invisible" to the storage facility and/or materials handling equipment. The storage rack or shelf, or the pallet or tote, occupy the same space whether full, partially full, or empty (except for nested empty totes and stacked empty pallets). The cube of the storage system is the basis for performance measurement.

For the purpose of explaining this concept, let us first discuss palletized storage in both pallet racks and floor stacked storage. We will discuss totes and shelf storage, and mechanized storage later in this chapter.

In general warehousing operations (whether rack, shelf, or floor stacked pallets), the maximum net cube of the storage space available for placement of merchandise in the warehouse is in the range of 30% of the total cube of the building area occupied by the "net operating space." The net operating space is defined in the same way as in manufacturing. It is the floor space actually used for storage operations and the internal supporting storage aisles within that space.

In a warehousing situation we also absorb the cost of supporting service space into the net operating space. If the warehouse is free-standing, this will be similar space to the service areas discussed in relation to manufacturing. If the warehousing operation is a part of a manufacturing plant, as in the case of receiving stores, work-in-process, or finished goods activities, the supporting services will be treated as a part of the manufacturing space's "use rate" computation. In that case the warehouse's net operating space will only have to absorb its own internal office support space and possibly toilets, lunch, and locker facilities.

In some cases, one may wish to separate the shipping and receiving areas and treat them as separate cost centers or functions. In this discussion, we will treat the shipping and receiving department areas and the main or "spine" aisle of the warehouse as a separate function or department for cost allocation purposes. The reason for this treatment will be addressed in the materials handling section of this chapter.

In the development of the space or "cube" "use rate" for warehousing operations, one starts with development of the same building or floor space "use rate" computation as in manufacturing. The space "use rate" includes all of the same period type space expenses, burden costs, overhead, and general and administrative (G & A) expenses, as in the manufacturing area. The space "use rate" is also the receptacle or "bucket" for collection of supervisory, maintenance, and administrative labor and material costs. The method for computation of this space "use rate" has been fully discussed in previous chapters. In some cases, the cost of the warehouse's building space may differ from the cost of the manufacturing space. This difference must be accommodated in the computation of the "use rate". At this point we will accept the computed operating space "use rate" for the warehouse as a "given" based on the same calculations as used in manufacturing.

However, in order to apply the storage and handling concept of "time use of cube utilization cost" or "cube handling cost," one must modify the basic "use rate" and apply it to the "cube" of the product in an equitable manner.

The storage area's space "use rate" can be modified to represent the planned net or designated space under the actual storage of goods (i.e., under the pallet stacks, pallet racks, or shelving), and the internal aisles' "use rate" expense can be absorbed into this actual or net storage space "use rate." This will have the same effect as the absorption of the facing aisles' space into the manufacturing work station's "use rate." The net storage space "footprint" will effectively be the area of the storage "work station" including one-half of its facing aisle in most cases (or the whole aisle if facing a wall or non-storage activity).

Since the cube of the available storage space is the logical medium for timed storage cost collection, it is necessary to convert the storage foot print space "use rate" into an available storage cube "use rate."

All of the available storage cube will usually not be fully occupied by goods. However, if "locked stock" is assumed to be forbidden, and items are not stacked on top of, or in front of other items to block access, the storage cube will be effectively occupied whenever a storage position has material in it. It cannot be used for other purposes if the pallet rack position, shelf, or floor stack position, is partially occupied.

The real or actual cube of the stored material would therefore appear to be charged with the whole "use rate" cost for completely occupying the cube of the standard storage module or position.

But the whole height of the storage pile in the footprint space or in the rack or shelf position is not occupied by the storage of material. Pallet stacks are not always full height. When racks and shelves are in use, there is normally a great deal of lost cube space even when the storage furniture is nominally full.

This lost cube includes the space occupied by the pallet, the clearance between the top of the load and the pallet and rail above, the flue between the

Warehousing and Handling Operations

backs of the back-to-back racks, and the "dead space" created by pallets which are nominally full but not cubically full. This cube occupancy situation is even more pronounced in shelving where small items occupy shelf space but do not usually fill the available cube.

Therefore, each pallet, rack, or shelf storage position represents a standard cube module. The cost of its use will have a "use rate" in the same manner as the cost of a work station.

In order to define this storage module, it is necessary to establish a nominal storage height for the storage footprint. In the case of floor stacked pallets, the standard storage height would be the total height of a standard cube or pallet load of the type of materials to be stored, not including the pallets, and assuming stacking of the materials in storage to the fully allowable storage height based on warehouse ceiling height and sprinkler clearance allowances (Figure 15.2).

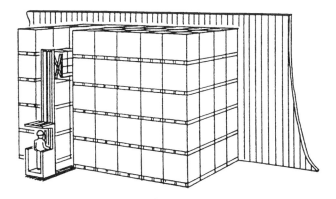

TYPICAL WAREHOUSE RENTAL COST PER ANNUM $1.50 PER SQUARE FOOT

OCCUPANCY FACTOR

AISLE LOSS   50%
SERVICE SPACE 20%
NET AVAILABLE FOR STACKING 30%
AT 100% OCCUPANCY
COST PER OCCUPIED SQUARE FOOT $5.357 PER ANNUM
STACK HEIGHT 20 FEET
COST PER OCCUPIED CUBIC FOOT $0.267 PER ANNUM
ASSUME 80% NORMAL PEAK OCCUPANCY
COST PER CUBIC FOOT OF INVENTORY $0.33 PER ANNUM
COST PER CUBIC FOOT PER MONTH $0.0275

Figure 15.2 Computing the bulk pallet storage cube cost factor.

In the case of rack storage, the height would also be the net height of material on pallets which could be stored in the racks, not including the pallets, racks, or clearances. For shelving, it is customary to define the net available space as 50 to 70% of the shelf cube depending upon the type of material being stored. These definitions are obviously arbitrary but they are adjustable to suit a real life situation in order to refine the precision of storage cost allocation.

The per square foot net storage space "use rate" will be divided by the standard or allowable storage height of material to establish a cube "use rate" per cubic foot of available storage cube or space. This rate will then be applied to the available cube for each rack position, shelf position, or pallet to establish a standard "use rate" for each storage position and method.

Each pallet, rack, or shelf position will then have a standard "use rate." This "use rate" will be charged to the stored product on a time use of facilities basis in the same manner as the work station "use rate" in manufacturing. As in the case of unused manufacturing capacity, the unused pallet positions' "use rate" costs will be treated as a warehouse expense variance and be charged to the profit and loss account as unabsorbed overhead or burden expense. Only the actually occupied space costs will be charged to the product cost accounts.

As an example of the application of this concept, let us assume a pallet load of cartons containing purchased parts. The pallet is a 48 × 40 inch standard unit and is stacked with four tiers of 12-inch high cartons. Thus the net cube of the material, assuming 1 inch of overhang on each side (42 inches) and no underhang, is 56 cubic feet.

The net storage space "use rate" for floor stacking is $0.72 per square foot, per day, and the net storage height is 36 feet. The "use rate" per cubic foot, per day is $0.02 per cubic foot, per day (24 hours). The storage cost per day for the pallet load is therefore 56 times $0.02 per day, or $1.12 per pallet, per day. This storage cost will be added to the daily cost of inventory insurance, personal property taxes, and interest on the money invested in the inventory, to compute the total cost of holding the material each day. This daily cost will be multiplied by the number of days the inventory supply is kept in the warehouse to calculate the stocking cost to be added to the manufacturing cost of the product. This storage cost will increase the landed cost of the product. This ownership cost is also an incentive for using a Just-in-Time philosophy of materials management.

The same computation procedure would be applied to the goods stored in pallet racks and shelving. However, in that case, the cubic space "use rate" must include the "use rate" computed for the storage furniture (racks and shelving). The storage furniture "use rate" would be computed in the same manner as has been described for manufacturing equipment. This storage equipment "use rate" would not include the pallets or tote boxes. They would have a

Warehousing and Handling Operations

separate "use rate" based on a shorter service life than that used for the racks and shelving.

In the case of raw materials, supply stores, purchased materials, and finished goods, the storage costs should be computed on the basis of a 24-hour day in the warehouse or storeroom. In some cases, where rapid movement or Just-in-Time operations are involved, it may be appropriate to compute the storage cost of work-in-process (WIP) on a single shift (or even an hourly) basis. This would be a judgment call by the cost system designer.

B. APPLYING THE "USE RATE" TO MATERIALS HANDLING EQUIPMENT

Materials handling equipment presents a complex cost assignment problem because of its variety and flexible application. Most materials handling equipment performs a variety of non-cyclic operations and interfaces with a variety of products during a typical work day. Those materials handling operations which can be identified as being dedicated to a single work station or process can usually be costed as a part of the work station. The roles of other materials handling equipment are more difficult to define.

For costing purposes, it is desirable to classify the various types of materials handling equipment on the basis of their level of mobility and the repetitiveness of their functions. These classes might be as follows:

1. Fixed and limited area handling equipment might include: Conveyors, feeders, elevators, stackers, jib cranes, hoists, monorails, bridge cranes, palletizers, etc.
2. Mobile materials handling equipment would include: Forklift trucks, pallet transporters, tractor-trailer trains, automated guided vehicles, front end loaders, trucks, people movers, etc.

In general, fixed and limited area materials handling equipment can be treated as being attached to the building structure. If we take this position, the costs generated by these devices can often be allocated to the net operating area and be handled in the same way as the air conditioning and heating equipment which is attached to the building.

In addition to the cost of ownership "use rate," some of the fixed and limited area equipment (such as a bridge crane) requires an assigned operator. The labor cost for the operator must be added to the ownership cost's "use rate" to develop a complete "use rate" for the equipment. In most cases, the cost of power and maintenance supplies can be included in the overall building "use rate" and ignored at the handling equipment level. Thus the "use rate" for the fixed and limited area equipment can include the combined costs of ownership, energy, and operator labor.

Fixed and limited area handling equipment is normally located in a specific department or in a well-defined geographic area of the plant. In a department location, the equipment's computed "use rate" can be divided by the total square foot area of the department and then be reassigned to each of the department's work stations in proportion to their individual areas. If the equipment serves a more general geographic area with several departments, the same approach can be applied on the basis of the whole area being served. Again, the equipment's "use rate" can be divided by the square footage of the area it serves and the charges reassigned to the individual work stations on the basis of their particular areas. The equipment's time use will be included in the time use or "use rate" charges of the work station. This cost will be applied to the product as a part of the work station's "use rate" cost.

Alternatively, this type of equipment can be isolated for costing purposes. It can be treated as a work station performing its own process operations in the production sequence. This approach might apply to a pallet loader, an automated empty pallet stacker, or the stacker tower of an automated storage/retrieval system. In this case, the cost must be applied to the product on the basis of units or cubic feet handled. This will require either manual or automated counting and posting of the volume handled into the production record. If an automatic identification and/or bar code system and direct computer input is in use, this might be a practical approach. However, if manual data collection is required, the cost effectiveness of this method is subject to question.

In general, the author recommends the area and work station assignment technique as the more equitable method of capturing the costs of this class of handling equipment.

Mobile materials handling equipment which is shared and not assigned to serve a particular operation or work station presents a different problem. In this case, the equipment may serve many work stations at random or be reassigned on a daily or even an hourly basis. There is no practical way to segment and assign the equipment cost to each of the operations it serves.

In addition, the cost of such equipment cannot always be represented by the "use rate" alone. In many cases the mobile equipment requires an operator and in some cases it consumes fuel. Most mobile equipment also requires both scheduled and unscheduled maintenance support which is not usually a part of the maintenance element of the facility's "use rate." The computation of the "use rate" for mobile equipment must include all of these cost elements.

Development of the ownership cost's "use rate," which will include the straight line depreciation or lease charges, interest, insurance, and taxes, can follow the same computation procedure as has been recommended for manufacturing equipment. If operator labor is required, the sum of payroll and fringe costs can be included in the equipment's "use rate." However, fuel, maintenance, and housing or space occupancy present a different problem.

Warehousing and Handling Operations

In most fuel-powered mobile handling equipment, the fuel consumption rate is predictable on the basis of engine running time and task configuration. The fuel use rate can be computed from manufacturer's data or it can be derived by averaging the recorded fuel usage. Fuel consumption, for a given task pattern, is usually fairly uniform over the life of the engine. Therefore, a fuel consumption standard is realistic and it can be included in the computation of the "use rate" with some degree of confidence. In the case of an electric battery driven unit, the power consumed by recharging the batteries can be measured for use as a basis for computing a "use rate" charge.

Conversely, the cost of lubricants and maintenance have a tendency to increase with the age of the equipment. The use of scheduled preventive maintenance and a regular lubrication program will not only add to the service life of the equipment, but it will also help in the predictability of these costs. This will allow the development of a maintenance and lubrication cost standard for the predicted service life of each machine.

The life-time standard will overabsorb these costs in the early years and balance the life time cost of the equipment by underabsorption in the later years. The establishment of an allowable maximum maintenance cost variance before triggering the overhaul or replacement of the equipment will provide a basis for monitoring the service condition of the materials handling machinery.

The combination of the lubrication, maintenance, and fuel or energy standards should be treated as an operating cost standard. This cost standard should be assigned to the equipment in combination with its ownership "use rate." The combined cost should be applied on a time use basis. The method for assigning time use will be discussed below.

Another part of the cost of mobile materials handling equipment is its "housing" cost. This cost can be ignored and the cost of the building and maintenance facility can be absorbed into the overall facility "use rate" and be distributed to the product via the work station charges. Conversely, the housing cost can be arbitrarily defined and added to the equipment "use rate" charge. If that is the desired course of action, the procedure should be as follows.

In order to establish a basis for housing general use mobile materials handling equipment, let us isolate the main "spine" aisles of the factory and/or warehouse and add the area of the receiving dock and shipping dock to the area of the spine aisles. The total of these common areas can be treated as the work station area of the mobile equipment fleet. The overall space "use rate" developed for the net operating area can be applied to these common areas. This space cost can be distributed to the primary mobile handling equipment by dividing it by the number of primary powered units (forklift trucks, pallet transporters, tractor-trailer trains, and/or automated guided vehicles). This assigned share of the space charge should be added to the equipment's ownership "use rate" and any operator charges to develop an overall "use rate" for each primary unit. The ownership "use rates" of non-powered handling equipment (pallet jacks and tools) should be added into the space "use rate" and distributed to the primary units. Figure 10.2 is a graphic model showing a computation method for developing the mobile materials handling equipment "use rate" with the inclusion of fixed and supporting materials handling equipment.

To demonstrate this type of calculation, let us use the following example. Let us assume that the main spine aisle is 15 feet wide and 100 feet long, the receiving dock is 60 feet wide and 50 feet deep, and the shipping dock is 120 feet wide and 50 feet deep; the housing space for mobile materials handling equipment would be a total of 10,500 square feet. If we use the space "use rate" quoted above for storage space and warehouse operations ($0.72 per square foot per 24 hour day), the daily cost of this space is $7,560 or $315 per hour. If there are five forklift trucks in the operation, the hourly housing space "use rate" per truck would be $63. Added to this would be the ownership cost of the non-powered handling equipment.

These figures raise a question concerning the reality of using this type of space cost allocation for mobile materials handling equipment. Alternatively, the spine aisle space cost can be absorbed into the net operating area space "use rate" and be allocated to the various production work stations. Treatment of the shipping and receiving dock space as separate work stations can be helpful in the management and evaluation of those operations if that is the desire of the management.

In both cases, the "use rates" of the mobile materials handling equipment can be computed without including a housing factor. The equipment time can be charged to the shipping, receiving, and production operations on a time use basis.

The time use of mobile materials handling equipment in a work station can be defined on the basis of a work cycle synthesis. This can be done by applying work study procedures and predetermined work standards or work sampling studies. The time use can also be the assigned work time, the operator's work time, or the records of a dispatch control system. Any attempt to capture the cost per pallet moved or the time required will involve a significant clerical effort which the author believes would be "overkill."

Absorption of the cost of general use materials handling equipment and personnel into the work station "use rates" appears to be more practical and cost effective. In general, the author feels that this is the most practical and realistic method for the application of mobile materials handling costs to the manufacturing operations which share the use of the equipment.

C. APPLICATION OF STORAGE AND HANDLING COST TO THE PRODUCT'S INVENTORY VALUE

In order to capture the cost of materials handling and storage, and to include these expenses in the cost of the manufacturing process, it is necessary to treat the storage and materials handling functions as operations in the manufacturing process.

In the case of a storage operation, this is fairly easy because the part or component must physically leave the prior manufacturing (or receiving) operation and travel to a specific storage location. The item is normally recorded out of the prior production operation and must be posted into the inventory of the warehouse or the storage location. When leaving raw material or WIP storage en route to the next manufacturing operation, or moving from the finished goods warehouse to the shipping department, the item's inventory transaction must again be recorded and the item must be moved.

In each of these moves, the date, and often the 24 hour transaction time is also recorded, either manually or by entry into the computer. This is particularly true when an automatic identification or bar code tracking system is in use. From this record of the in and out movements or transactions for each item, one can define and compute the item's dwell time in storage. The dwell time can be computed on a daily, shift, half day, or hourly basis, depending upon the level of precision desired.

This computed or actual storage time should then be multiplied by the time "use rate" per cubic foot of usable storage capacity as defined above. The result of this calculation should then be multiplied by the cubic feet of the product stored in the transaction lot or in the product's stored inventory to produce the time cost of storage for the time the item is held in inventory. If there is a scheduled holding time for the item to reside in storage, this should be compared with the actual recorded time to define the storage cost variance. The standard and actual storage cost should be added to the accumulated level-by-level inventory value and/or actual cost of the item. The variance should be reported to management and its cost should be entered into the profit and loss account.

The capture of generalized mobile materials handling equipment and materials handling labor costs can be more difficult. In the case of materials handling operations which are dedicated to a particular work station or operation, the equipment's "use rate" and the operator or manual handling labor costs should be treated as elements of the cost of the operations being performed by that particular work station.

The materials handling equipment's time can be defined by using the operator's clock time, by putting a timer on the vehicle, by using a time based fuel consumption rate factor, or by using predetermined time standards and synthesizing the operation's time. The handling operation's standard cost should be added to the level-by-level product cost generated in the work station served and it should be entered into the inventory valuation. Any computed variances based on actual time versus standards should be treated in the same manner as manufacturing variances. They should be reported to management and recorded in the profit and loss accounts.

However, in the case of a multiwork station handling support system, or plantwide mobile materials handling systems using vehicles such as forklift trucks, automated guided vehicles, or tractor trailer trains to serve several areas or the whole plant, allocation of the cost of each handling operation to a particular product is nearly impossible. These systems normally haul and handle a variety of products during a typical work day and are often in an idle "waiting" mode.

One can easily define the time based or hourly cost of using these systems by applying the above described computation methods to calculate their "use rates" and operator labor costs. These computations will ignore the "housing" factor. They will include any fuel, insurance, taxes, maintenance, and other ownership costs of the equipment plus the cost of the operator if one is required.

This computation will provide a time based "use rate" for the equipment and operator labor of the generalized materials handling system. The labor cost of general manual labor "gangs" can be treated in the same manner.

This cost will be considered as a manufacturing burden item and will be applied to the cost of the operation via the space "use rate" and in proportion to the system's *scheduled* hours of use.

In order to absorb a fair share of this generalized materials handling operation's expense, it will be treated as a manufacturing *burden* item. The total cost of the system will be divided by the *scheduled* hours of plant operation. The resulting hourly "use rate" will be added to the space "use rate" of the plant. In this manner, a fair share of this generalized materials handling cost can be applied to each work station in proportion to the work station's area. The generalized materials handling cost will therefore be a component of the work station's "use rate." It will find its way into the product cost via the product's time use of the manufacturing work station. As in the case of dedicated materials handling operations, any computed variances based on actual time versus standards should be a part of the manufacturing variances. They should be reported to management and recorded in the profit and loss accounts.

D. DISTRIBUTION CENTER/WAREHOUSE COSTING

When a manufacturing company chooses to have a separate and freestanding distribution center for finished goods, or a wholesale or retail enterprise oper-

Warehousing and Handling Operations

ating out of a distribution center, a different approach to storage and materials handling cost is applicable. In a distribution center, the materials handling cost is the *direct labor* of the operation. With rare exceptions, as when packaging and assembly operations are included, the distribution center adds no value to the product. It is an all cost operation! However, the distribution center is also a profit or cost center. In many cases, as in the public warehouse industry, it is the whole enterprise.

Therefore, one must address the cost of operations in a distribution center with care and precision, but from a different point of view. As stated above, the cost of storage and materials handling is proportional to the cube handled and has no relationship to the identity or value of the product handled. Factors such as fragility, security, shelf life, weight, and shape have an impact on the design of the handling and storage system. They also have an effect on the amount of inventory which is held in the warehouse and the length of time it is kept. However, once the system has been designed for a population of products, the system's "common denominators" make the product invisible to the system and the resulting operating costs are strictly unit and cube based.

The distribution center uses space, storage equipment, materials handling machinery, and other types of fixed equipment such as packaging, palletizing, stretch wrapping, and automated storage and picking systems. In this sense, its capital facility cost elements are identical to those of a manufacturing enterprise. The development of "use rates" for allocation of the costs of equipment, facility ownership, and general operations, can follow the same procedures as were recommended for the manufacturing operations.

Referring back to the computation of manufacturing cost standards and "use rates," the treatment of the basic facility or space costs, burden, overhead, general and administrative expense, and other period costs in a distribution center would be handled in the same manner as in manufacturing. In addition, the ownership cost of storage furniture such as pallet racks, cantilever racks, shelving, frame pallets, pallets, totes, and other storage equipment would be treated in the same way as manufacturing equipment and tooling. "Use rates" will be developed for each item and their costs will be included in the overall "use rates" of the work stations where they are located.

Specialized warehousing machinery such as carousels, automated storage/retrieval machines, automated order filling machinery, automatic pallet stackers, automatic pallet loaders and unloaders, packaging machinery, conveyors, sorters, and computers will be treated in the same manner as production machinery. In each case an appropriate "use rate" will be computed for the equipment and it will be assigned to a work station.

It is in the definition of the work stations and in the assignment of general materials handling and other operating costs that the distribution center receives slightly different treatment. While most of the labor and equipment costs can be assigned to a definable work station, some labor and mobile equipment will be shared and serve multiple work stations. For this reason, some of the work stations are without space assignments and are treated as general burden cost generators.

In the distribution center, the most common work station breakdown might include such areas as:

- Receiving
- Inbound Inspection and Checking
- Packaging and Kitting
- Inbound Vehicular Transportation to Storage
- Bulk or Reserve Storage Operations
- Pallet Rack Storage Operations
- Manual or Vehicular Case Picking
- Automated Storage/Retrieval Systems
- Manual or Vehicular Open Stock Order Picking
- Automated Sortation Systems
- Outbound Vehicular Transportation to Shipping
- Conveyor Systems
- Order Inspection, Packing, and Addressing
- Unitizing and Stretch or Shrink Wrapping
- Shipping and Loading
- Order Processing and Documentation

In some of these work stations, notably the conveyors and automated systems, no operator labor is required. In some instances, as with the conveyors and sorters, there is little definable floor space because the equipment and the operation is often overhead and above other more identifiable work stations. In such situations, the equipment's "use rate" must stand alone without any added labor or space elements in the work station's total "use rate." The "use rate" charges for such work stations will be distributed to the product via the net operating area of the distribution center and will be absorbed into the "use rates" of the work centers through their space charges.

Another group of work stations which will receive the same treatment includes the transportation or movement from receiving to storage, from one level of storage to the subsequent level (i.e., from reserve to case pick, or from case pick to open stock, etc.), and from packing to shipping. These costs will include the equipment "use rates" and any labor costs, but no space charges. These "use rates" can be distributed into the product cost via the net operating area's "use rate" and its space distribution to the various work stations, or they can be assigned to the product on a unit handled basis.

In most of the other work stations, there will be a definable allocation of space and assigned labor.

Warehousing and Handling Operations

In some cases, as in packaging, packing, stretch or shrink wrapping, there will be some consumption of "production" materials which add cost or "value" to the shipped product. The value added by packaging is a component of the product's function in its movement through the system. It is a part of the product as sold to the consumer. It protects the product and in many cases displays and sells the product, regardless of its location. This expense adds to the landed cost of sales of the product.

However, the author does not accept the concept of "time and place utility value" being added as a part of the value system in defining the landed cost or value of a product.

Packing is the act of combining several units or items for shipment. This is a part of the distribution cost and is not a component of the product's manufacturing cost. However, it also adds to the "landed cost of sales." All of the transportation and distribution operations generate an expense cost but add no value to the product!

It is through definable work stations that one can capture the distribution center's cost and apply it to the product. This can be done on a cost per unit handled basis by tracking the order processing system's records. Once the standard throughput cost has been defined for each item passing through the system, the standard cost can be applied to the inventory record on a line item basis as each transaction is posted.

For example, the cost of storage can be charged to the value of the item at the time it is drawn from stock and its time in storage can be defined. The standard handling and order picking cost can be posted to the inventory line item's value for each receiving report and outbound order picking transaction.

However, a key factor in this type of operation is that the handling characteristics of the materials and their "common denominators" change as they flow through the work centers. For example, in the reserve stock area, most of the handling might be on pallets using forklift trucks. In the case picking area, the inbound movement would be in pallet form with forklift trucks or turret trucks, while the outbound flow would be in man handleable cases, possibly using high order picking vehicles. Likewise, in the open stock picking area, the inbound flow would be in cases and the picking would be in individual units or packages. These changes complicate the costing of the operations in the same way that the steps in a manufacturing work center affect the costs of those operations.

An operations sheet with operations' time standards is required in manufacturing to define the process steps for each part and product. This type of documentation is also desirable to precisely describe the procedures and define the costs of the handling operations in a complex distribution system.

However, as discussed above, the costs generated by materials handling and storage operations relate to the cubic feet and handling characteristics of the products being handled and stored, and not to their identity or value. Also, in view of the complexity of the product mix and non-cyclic operations found in many distribution centers, assignment of costs to individual products must be on a cube and handling class basis. The methods or operations sheet should address the product groups on a generalized group technology basis using handling class and handling or storage functions.

For example, order picking is an operational function which can apply to several handling classes. Each would have a different set of cost standards. The products would be classified by handling class and the standard costs would be developed for each handling class and each operation or function.

This requires the establishment of definable handling classes. These classifications must be assigned to products so that they can accumulate the correct standard costs as they progress through the operational stages of the distribution center.

This approach is similar in concept to the use of a part's geometric characteristics as the basic criteria for processing decisions in a manufacturing group technology system. Typical handling classes are:

- Finger handleable—screws, buttons, diodes.
- Hand handleable—tools, dishes, clothes, health and beauty aids, small parts.
- Two hand handleable—cases of canned food, grain sacks, camp stoves, tote boxes.
- Two person handleable—100 pound sacks, tables, crates, mattresses, sofas.
- Machine handleable—pallet loads, machinery, shop tubs of parts, pool tables.
- Long light goods—fishing rods, skis, plastic tubing, aluminum extrusions.
- Long heavy goods—lumber, pipe, steel plate.

To demonstrate these handling and storage cost application concepts, we will use only a few of the product handling classifications and apply them to a conventional distribution center operation. Let us consider a typical grocery and health and beauty aids distribution center which services a chain of supermarkets with dry groceries and health and beauty aids. The distribution center receives all of its goods by highway truck in fully palletized loads. All of the grocery products are shipped to the stores in single and multi-case lots or as open stock in cartons or tote boxes. A few go in full pallet loads. All of the stores usually have hand pallet jacks.

All of the health and beauty aids are shipped to the stores in either single case quantities or as open stock. The open stock is placed in non-returnable scrap cartons which are baled at the stores or in returnable totes. Mixed goods

Warehousing and Handling Operations

from the full case and open stock picking operations are palletized and stretch wrapped before being loaded onto the trailer to the stores.

All case order picking is performed on high order picking vehicles. Open stock picking is from manual slide through, rear loaded, picking shelves. Open stock picking is normally into cartons or totes and often onto a conveyor leading to the order accumulation operation at the shipping dock.

The case picking rate is 10 cases per minute and the open stock picking rate is 20 pieces per minute. It takes ½ minute to place or pick up a pallet and the vehicles move at an average of 200 feet per minute including starting and stopping. Truck loading and/or unloading requires 30 minutes for 20 pallets per load. Stretch wrapping requires 1 minute and it takes 3 minutes to assemble a mixed load pallet.

To demonstrate the cost system application, we will address the pallet rack storage and case picking module or work station and its operations and the open stock picking operation and its work station. We will make some assumptions and use some "given" cost data. The assigned "use rates" are assumed to have been computed by using the same procedures as were described above for the manufacturing operations. The flow of goods will be based on the receipt of a pallet load of mouthwash and the picking and dispatch of the mouthwash to a retail store in a carton or tote box.

Let us assume a construction cost of $50 per square foot, a net operating space factor of 60%, and a 30 year straight line depreciation.

Thus: $50/0.60 = $83.00 per net operating sq. ft.

Applying the thirty year depreciation and the 2000 hour work year, this results in a computed space "use rate" of $0.0014 per square foot, per hour, or $0.0112 per square foot, per 8 hour day. Although the goods are normally in storage for 24 hours, the 8 hour charge will still apply because the plant operates on a 2000 hour schedule.
Thus

$83/30 = $ 2.77 and $ 2.77/2000 = $ 0.0014 per hour

and:

$0.0014 × 8 = $0.0112/sq ft, per 8 hour (or 24 hour) day
for depreciation

Let us assume that the total of all of the elements of burden, overhead, and period costs is equal to the building depreciation cost of $50/30 or $1.67 per square foot, per year or $0.00084 per hour, or $0.0067 per 8 hour (or 24 hour) day. Therefore:

$0.0112 + $0.0067 = $0.0179 per hour

This makes the space "use rate" equal to $0.018 per square foot, per hour or $0.144 per square foot, per 8 or 24 hour day.

Let us assume a maximum merchandise stack height of 20 feet. This gives us $0.144/20 = $ 0.0072 per cubic foot per day.

This is a net cube "use rate" of $0.0072 per cubic foot, per 24 hour day. This cube "use rate" includes the space "use rate" and all of the period and indirect expenses of the distribution center for floor stacked product.

With a nominal or standard pallet cube capacity of 64 cubic feet of merchandise, this will result in a daily (24 hour) cube or occupancy "use rate" of $0.461 per day for rack storage of a pallet of goods. In addition, let us assume a rack "use rate" of $0.06 per day, per pallet rack position, for the pallet rack equipment. This will result in a standard daily storage cost of $0.521 per pallet racked pallet at a nominal 64 cubic feet per pallet.

Dividing by 64 cubic feet, this results in a standard cube "use rate" of $0.008 per cubic foot per day for palletized rack goods.

If there are thirty-six 1½ cubic foot cartons per pallet, the standard daily storage space or cube cost will be 36 × 1.5 × $0.008 per cubic foot ($0.012 per carton) = $ 0.432 per pallet, per day for racked and palletized goods on a cube utilization basis.

Thus the pallet position cube is 84% occupied and the empty pallet position space costs management $0.08 per day as a distribution center overhead cost.

The time use of facility system uses the actual cube utilization to develop operating costs for assignment to products. The standard cost is assigned to the product via the inventory posting on a handling unit cost basis. This is calculated from the cubic feet of product in the "common denominator." This generates a standard versus actual cube utilization variance for over or underabsorption of overhead. The cube utilization variance is charged to the distribution center's operating statement or its profit and loss account.

Applying this same standard cube "use rate" to the flow through pick racks, and assuming 1½ cubic feet per carton, the storage cost of a carton of goods will also be $0.012 per 24 hour day. If the "use rate" of the flowthrough rack, for 10 cartons per slide, is $0.03 per slide, per day, the total storage cost will be:

$0.03/10 + $ 0.012 = $ 0.015 per carton per day

Assuming a pallet pattern of six cartons per tier and a transfer pattern of one tier per restock cycle to the flow racks, the average picking inventory is assumed to be seven cartons for an average pick line storage cost of $0.015 × 7 = $0.105 per day, per slot.

Thus, the average daily storage cost for one racked pallet plus the pick line inventory of this item would be the pallet storage cost of $0.432 plus the pick line storage cost of $0.105 per slot, or $0.537 per day. At 36 cases per pallet and seven cases average pick line stock (43 cartons picking stock), this is $0.0125 per carton, per day.

This same calculation process, adjusted to carton size, will apply to every carton item in the distribution center which is picked in both case pick and open stock. Differently shaped goods which are handled and stored in another manner would have different storage costs.

In terms of the handling costs, the movement from the inbound truck to the pallet racks would require a tailgate pick-up by the forklift truck, an assumed 200 foot travel to the rack, placement into the pallet rack, and an empty 200 foot return travel. If the "use rate" for the forklift truck is $30.00 per hour or $0.50 per minute, including all ownership, fuel, maintenance and overhead costs, each 30 second pick-up and its subsequent rack placement of a pallet would cost $0.50 and the round trip move would cost another $1.00 for the vehicle. The total vehicle cost would be $1.50 per pallet trip.

If the operator is paid $10.00 per hour and the fringes are 35%, the 2 minute operator cost for the move would be $0.45 including the return trip to the truck dock. This would result in a total inbound move cost of $1.95 per 36 case pallet, or $0.054 per carton.

Assume that the case picking and open stock picking vehicles have "use rates" of $40.00 per hour or 0.667 per minute, and the order picking refill location is 100 feet from the pallet rack area. If the case picking rate is ten cases per minute, the cost of transferring the six cartons to the pick rack would be computed as follows.

The vehicle requires one minute for the round trip from the pallet racks to the pick racks and ½ minute for loading the slide. The picking vehicle requires 0.6 minutes for the picker to select six cartons for a total of 2.1 vehicle minutes. At $0.667 per minute, this is a total of $1.40 in vehicle cost.

If the operator is paid $10.00 per hour with 35% in fringes, this results in a picking labor cost of $0.225 per minute. The selection of six cartons or cases will require 0.6 minutes at a cost of $0.135 for picking labor. Adding the vehicle cost of $1.40 for the operation, the total cost of moving and picking six cartons is $1.535 for a cost of $0.256 per carton for case picking and picking stock replenishment.

The manual open stock picking rate is 20 piece picks per minute. If we assume 24 pieces per carton, and a picker pay rate of $9.50 per hour plus 35% fringes($12.825 per hour), the picking labor cost (@ $0.214 per minute) will be $ 0.0107 per piece. Mathematically presented, this is:

Labor = ($9.50 × 1.35)/60 = $0.2137/Min, or $0.214/min
$0.214/20 = $0.0107 per picked piece
$0.0107 × 24 = $0.2568, or $0.257 per carton for picking

In summary, the cost elements are

Inbound pallet movement from truck = $0.054/carton
Rack and pick slot storage per day = $0.0125/carton
Transfer storage to picking = $0.256/carton
Open stock order picking = $0.257/carton

This results in a total throughput cost of $0.5795, or $0.580 per carton.

If it is assumed that the goods are in storage for only one day, the total cost of moving a carton from the inbound truck dock through the order picking operation into a shipper carton or tote box will be $0.580 per carton. This cost, plus the movement to the shipping dock and the carrier cost to the store, must be added to the manufacturing cost in order to compute the landed cost of sales for the product.

To demonstrate the impact of these seemingly small cost elements, let us note that a 40 foot highway van can hold approximately 1600 cases of grocery or health and beauty aids products at 1.5 cubic feet per case. Assuming that a single supermarket often uses up to four van loads per week and a distribution center is serving 40 supermarkets, it will move approximately 51,200 cases per day. At 24 pieces per case this is 1,228,800 pieces per day. In a 5 day week this is 256,000 cartons or 6,144,000 pieces per week. For this volume, the inbound move, the carton move to the pick line, plus the open stock order picking, accrues a one day turnover cost of $29,696 per day or $148,480 per week. This is only a segment of the distribution center's operation.

An additional $0.0125 cost per carton must also be added for storage for each additional day that the inventory resides in the warehouse. At 51,200 cartons per day, this is another $640.00 per day. This is a small cost factor for high value products but it can be very significant for low dollar density items. This cost is an added incentive to achieve rapid turnover or Just-in-Time operations.

These same types of calculations can be applied to bulk or floor storage, the packing and checking of the order, the palletizing and stretch wrapping, and movement to the outbound truck. The total cost of the movement through the distribution center should be calculated on an item, carton, pallet, day, and order basis.

The productivity of the workers can be cost evaluated and the profitability of each customer can be measured. All of these measurements of performance can be precise if the time use of facility "use rate" approach and the total absorption standard cost method is applied.

In the retailing, wholesaling, and public warehousing industries, the distribution center operation is the mainstream of the enterprise's business. The measurement and control of the costs of distribution operations is critical to the success of the business.

Warehousing and Handling Operations

In manufacturing, the warehousing of inbound materials, supplies, work-in-process, and finished goods should be treated as separate departments, cost centers, or work stations. Each handling and storage operation, at each stage of the logistical flow through the manufacturing enterprise, should be treated as a separate warehouse and its cost should be precisely measured and controlled.

A distribution center is a cost or profit center and a management unit of the enterprise. It must be measured and controlled with the same level of precision as the manufacturing operation. In addition, its role is an all cost operation and it usually adds no value to the product.

16

Sample Product Cost Calculations

A. THE PRODUCT AND THE BILL OF MATERIALS

In order to realistically demonstrate the "time use of facilities approach" and the "total absorption standard cost method" of precision manufacturing costing, it is necessary to apply the procedure to a real product. In this chapter, we have chosen to use a simple worm gear speed reducer as the model for cost development (Figure 16.1). We will compute typical cost elements as demonstrations of the total absorption, time use of facilities costing method. These elements will include computation of the standard "use rate" cost for applying indirect expenses to the product, the development of a standard labor cost for an employee, and the calculation of standard materials costs for parts. We will assume or estimate other cost elements in order to limit the volume of calculations in this chapter.

The author has applied these costing techniques and coding procedures for clients using both manual and mechanized systems. These procedures can be supported by either PC computers or mainframe equipment. In most manufacturing costing and materials management systems, the cost calculations can be computerized by using standard spreadsheet software or by writing customized programs. These precision costing techniques were tested for PC computer compatibility by using the FoxPro software, by Anas Alsawaf in his 1992 Master of Science Thesis in Industrial and Systems Engineering at Ohio University in Athens, Ohio. His thesis is in the Alden Library at Ohio University.

For every product, the first step in the application of the precision costing system is to analyze and define the parts of the product to be costed. This

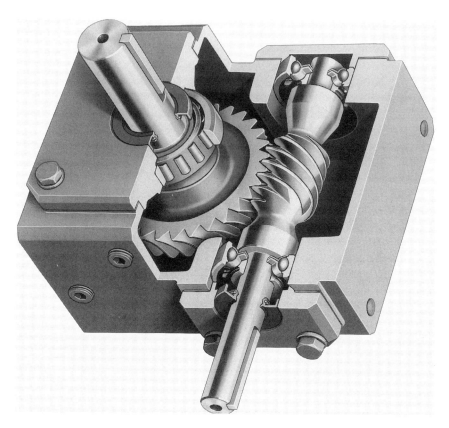

Figure 16.1 Typical worm type speed reducer. (Photograph courtesy of the Falk Corporation.)

requires the development of a numerical manufacturing bill of materials (MBoM). This MBoM should also include a level-by-level set of matrix type part numbers. Some products have multiple levels of semi-finished parts in the work-in-process inventory. Complex products usually require the breakdown of the completed product into multi-level sub-assemblies, detailed parts, semi-finished parts, purchased parts, and raw materials. In very complex products such as machinery, automobiles, aircraft, computers, etc., there are usually a number of sub-assembly levels and thousands of parts. In such cases, many of the parts may also be used in a multiple of sub-assemblies and at different indent levels in the assembly structure. Regardless of product complexity, the total absorption, time use of facilities, costing approach will be applicable to each part, component, and sub-assembly of the product.

Sample Product Cost Calculations

To demonstrate the process, it is best to use a simple product and to abbreviate the procedure. By using a worm type speed reducer (Figure 16.1), we have selected a product with parts that require multiple machining operations and several different manufacturing processes. This product also includes both parts and sub-assemblies, several different materials, and several purchased components. In addition, this product requires two distinct types of manufacturing processes (casting and machining) and a complex assembly process.

The speed reducer requires the manufacture of a housing composed of several machined castings, a one piece steel worm and shaft, a steel worm wheel (gear) shaft, and a bronze worm wheel. It also involves the purchase of four anti-friction bearings, two bearing seals, twelve hex head bolts, two Allen socket plugs, and three straight shaft keys. The numerical manufacturing bill of materials for this speed reducer would be approximately as follows:

```
Finished Product
    Worm gear speed reducer    01-001-0001
Major Sub-Assemblies           02-XXX-XXXX
    Housing                    02-002-0002
    One piece worm shaft       02-003-0011
    Worm wheel assembly        02-004-0004
Finished Parts                 03-XXX-XXXX
    Housing                    03-002-0005
    Front cover plate          03-006-0006
    Solid end plate            03-006-0007
    Seal hole end plate        03-006-0008
    Bronze worm wheel          03-009-0009
    Worm wheel shaft           03-010-0010
    One piece worm shaft       03-003-0011
    Two worm wheel spacers     03-012-0012
Purchased Parts                04-XXX-XXXX
    Two ball bearings          04-013-0013
    Two roller bearings        04-013-0014
    Two bearing seals          04-014-0015
    16 Hex head bolts          04-015-0016
    Two allen socket plugs     04-015-0017
    Three ¼" sq. keys          04-016-0018
Semi-Finished Parts            05-XXX-XXXX
    Housing casting            05-017-0019
    Front cover casting        05-017-0020
    Two solid end castings     05-017-0021
    Seal hole casting          05-017-0022
    Bronze gear casting        05-018-0023
```

Raw Materials	06-XXX-XXXX
Scrap iron	06-024-0024
Bronze ingots	06-025-0025
1045 CRS steel 3″ rod	06-026-0026
1045 CRS steel 2″ rod	06-026-0027

The size of this speed reducer will, of course, affect the selection of the machinery used to manufacture it and its parts. However, in this demonstration of the cost system, the dimensions of the parts and product need only be approximated. For the purposes of this chapter, we will assume that the cast iron housing is approximately 12 inches wide, 6 inches deep, and 14 inches high. We will also assume that the bronze worm wheel has a pitch diameter of 8 inches. This would make the outside diameter of the steel worm shaft approximately 3 inches. These dimensions may not be realistic from a design point of view and are not intended to be so. These dimensions are only meant to give the reader a mental image of the size of the product. The purpose of this chapter is to demonstrate the cost calculation procedures; the choice of the product is incidental. It is not critical to the costing procedure demonstration.

In this demonstration, we will assume a significant production volume (i.e., 1000–5000 per month) requiring a large, vertically integrated manufacturing plant and foundry, with modern numerically controlled machinery and large batches of worm wheel speed reducers on each manufacturing order. However, these costing techniques are equally applicable in a Just-in-Time environment which produces in batches of one unit.

B. THE MAKE OR BUY DECISION

The decision to make or buy a component or a part of a product is based on a combination of criteria. In cases where the capability and capacity to make the part or component is available within the enterprise, the decision to buy the item from a vendor can often be entirely a matter of unit cost or price. However, in most cases the decision to make or buy an item is more complex.

Technical capability, or the lack of it, is a primary reason to buy an item. If the component is technically specialized, as in the case of anti-friction bearings, it is obvious that a specialized vendor would be selected. Specialized commodity type items such as bolts, nuts, washers, etc. are also normally purchased from vendors rather than being manufactured in-house.

Another factor in the decision is the availability of the required facilities, skills, or capacity to make the components. For example, a foundry may have the ability and facilities to pour iron and steel castings but be unable to pour bronze or aluminum. In such a case, an expected low volume of non-ferrous castings may make the required equipment investment and skills development uneconomical. This situation would lead to the outside purchase of the nonferrous castings.

The same logic may apply to the machining of very close tolerance parts, or parts which are too large for the plant's available equipment. Many manufacturers buy outside support to supplement their in-house capacity or capability. They also often buy these services to avoid the purchase of large size equipment which would become an underutilized capital investment.

Another reason for outsourcing is the accommodation of irregular schedules. It may be possible to manufacture every component of a product in-house, most of the time. However, during surge seasons, or when unusually large orders are received, it may be necessary to add outside capacity to meet customer delivery schedules. In some cases, manufacturers outsource some production on a regular basis to maintain active and friendly vendors who will support such varying demands.

Quality control can also be a reason for outsourcing. If a vendor has superior skills or equipment, the ability to maintain quality may require such support. It is also possible to use such sources as a basis for comparison with, and measurement of, in-house quality and cost performance.

The costing of purchased components can be a simple matter of comparing prices. However, it is also wise to use precision costing techniques to estimate the vendors' costs and to evaluate their prices. Due consideration of the vendor's need to make a profit must be included in the cost analysis. In-house net costs do not usually include a profit computation and are not a true measure of a fair vendor price. In this demonstration case we will use actual or estimated vendor prices to cost the purchased parts, components, and materials. The price quotations which we will use are:

Ball bearings	$23.50 each
Tapered roller bearings	$13.00 each
Bearing seals	$ 2.00 each
Hex head bolts	$ 0.60 each
Allen socket plugs	$ 0.40 each
¼-inch square keys	$ 0.50 each
Scrap iron	$ 0.50 per pound
Bronze ingots	$ 2.00 per pound
1045 CRS rod	$ 0.70 per pound

C. CASTINGS AND FOUNDRY OPERATIONS

In order to manufacture the sample speed reducer in a vertically integrated plant, it will be necessary to have a foundry which is capable of casting both grey iron and bronze. The development of detailed costs for such a foundry would require the computation of "use rates" for two types of electric furnaces, manual or automated casting equipment, separate pouring, shake out and shot cleaning machinery, sand handling and core making facilities, and scrap iron and bronze ingot handling facilities.

Although computation of such cost standards are a part of the design of the cost system, their inclusion in this chapter would make it much too long and also unnecessarily complex as a means of demonstrating the operation of the cost system. In order to simplify the system demonstration, we will assume that the castings are delivered into the machine shop from an adjacent company operated foundry at a computed standard cost for each item.

Computation of the foundry space "use rate" and the "use rates" for each work station and machine in the foundry would be required along with the costs of the raw materials, contingent materials, and labor in order to develop these casting costs.

The castings will be delivered into the machine shop's work-in-process store room in shop tubs by forklift trucks. The tubs will be four feet square and will contain the following quantities. The cleaned and fully absorbed casting cost of each piece will be assumed to be:

Housing casting	20/tub @ 20 lb	(05-017-0019)	$14.00 ea
Front cover	60/tub @ 7 lb	(05-017-0020)	4.90 ea
Solid end	60/tub @ 8 lb	(05-017-0021)	5.60 ea
Seal hole	60/tub @ 8 lb	(05-017-0022)	5.60 ea
Bronze gear	20/tub @ 44 lb	(05-018-0023)	88.00 ea

The machining of these castings will be performed in the manufacturing facility's machine shop. Some of these machining operations will be used to demonstrate the cost system. However, some of the machining costs will also be estimated or assumed in order to limit the length of this chapter.

D. COMPUTING MATERIALS COSTS

The raw material cost is required as a component of the cost computation. The estimated costs of the castings were based upon actual quoted wholesale metal warehouse prices and the estimated weights of the parts. Steel warehouse price quotations have also been used for the 1045 Cold Rolled Steel (CRS) shaft stock.

The "nest and gain" factor and the chip, scrap, and shrinkage effects are also a part of the material cost computation. "Chip" losses from machining the castings and other machined parts are already included in the materials' standards. Their cost is included in the adjusted material cost estimate. The standard material cost includes the original purchase of these chips.

Sample Product Cost Calculations

In this example case, the chips and drop off could probably be recycled through the casting process and thereby slightly reduce the net cost of the casting materials. This could be considered a "nest and gain" cost saving but it is probably insignificant. The amount received for any salvaged chips which are sold back to the vendor is considered "miscellaneous income." It should be included in the profit and loss statement.

In the case of metal materials, the shrinkage factor is a function of inventory control and corrosion. In this example let us assume a shrinkage of 1% for the 1045 steel bar stock and 0% for the scrap steel and bronze ingots.

Scrap iron is being used for casting the housing parts. The cost of scrap iron is assumed to be approximately $0.50 per pound or $1,000 per ton. Consumption of this scrap material will not be significantly affected by "nest and gain" or shrinkage factors. Any dropoff, risers, flash, or spill from the casting operation, or chips from machining the castings, can be recycled back into the furnace and remelted for a later pour. The weight of each of the steel castings was estimated on the basis of 490 pounds per cubic foot of steel or 0.284 pounds per cubic inch.

The bronze ingots are assumed to be 5 inches square and 8 inches long. The ingot bronze is expected to weigh 0.322 pounds per cubic inch or approximately 556 pounds per cubic foot. Each ingot will weigh 64 pounds at $2.00 per pound, and cost $128 per ingot. Again, this material can be recycled and the chip and shrinkage losses minimized.

The 3 inch diameter, 1045 CRS steel rod will cost approximately $0.70 per pound. It is assumed to weigh 24 pounds per foot and cost $16.80 per foot of rod. The 2 inch diameter rod of the same material will weigh 10.7 pounds per foot and cost $7.49 per foot. The 1045 CRS steel rod will be delivered in 22 foot bar lengths. Chips and scrap from this material can also be recycled back into the foundry.

The 3 inch diameter worm shaft will have a standard material (M_D) slug length of 16 inches and will weigh 32 pounds for a value of $22.40 per slug. It will be finished to a length of 15 inches. The 2 inch diameter worm wheel shaft will have a standard material requirement of 14 inches or 12.48 pounds at a cost of $8.74 per piece. It will be finished to 13 inches in length.

In this sample calculation, we will compute a "nest and gain" factor for the machining of the worm shaft and worm wheel shaft. Let us assume that these parts will be made on an automatic CNC turn/mill lathe with a 3.5 inch diameter spindle hole and an automated bar feed system. The parts will not be made from precut standard slugs. Figure 8.1 compares the machining methods.

The "nest and gain" factor generated by using the automated system will be computed as follows:

Worm Shaft Stock, 3 inch diameter, 1045 CRS rod in 22 foot bars
Standard material slug length = 16 inches
Finished part length = 15 inches
22 feet × 12 inches = 264 inches per bar
16.5 standard slugs or parts per 22 foot bar
Cut-off tool = 0.032 inches thick
Start bar facing cut = 0.015 inches
Therefore, 15.047 inches per piece

Using Precut Standard "Slugs"
264 inches/16 inches per slug = 16.5, or 16 slugs per 22 foot bar
264 inches − (15.047 inches × 16) = 264 inches − 240.752 inches = 23.248 inches loss per 22 foot bar including 0.953 inches chucking stock per slug
23.248 inches/12 = 1.937 feet
At $16.80/foot this is a $32.54 stock loss per 22 foot bar, including chucking stock for each slug

Using a Bar Fed Lathe and Process Cut-Off
264 inches/15.047 inches = 17.545, or 17 parts per 22 foot bar
17 × 15.047 inches = 255.799 inches
264 inches − 255.799 = 8.201 inches chucking stock per bar
8.201 inches/12 = 0.683 foot per 22 foot Bar.
At $16.80/foot this is a $11.47 stock loss per 22 foot Bar

Therefore, a 22 foot bar would produce 16 worm shafts from precut standard material slugs, or 17 worm shafts if the operation is performed on a bar feed machine. If the standard 16 inch material slug is used, there would be a 23.248 inch drop-off (D/O) from the bar. This would be a part of the chip loss of the operation and its cost would be distributed to the 16 parts at 1.453 inches per part and be included in the purchase price of the 16 inch slug and the 22 foot bar.

If the operation is performed on a bar feed machine, the 17 pieces made from the 22 foot bar would produce a drop-off of 8.201 inches. The total material "nest" saving would be

(23.248 − 8.201 inches) = 15.047 inches per bar,
or one additional worm shaft per bar

This will result in the production of one additional 15 inch worm shaft part per bar. This material saving is a "nest and gain" factor (F_{ng}) of 5.69% for this part based on the use of the bar feed manufacturing method instead of the standard design slug material allowance.

Sample Product Cost Calculations

Worm Wheel Shaft Stock, 2 inch diameter, 1045 CRS rod, 22 foot bars
Standard material slug length = 14 inches
Finished part length = 13 inches
22 feet × 12 inches = 264 inches/bar
18.85 standard slugs or parts per 22 foot bar
Cut-off tool = 0.032 inches thick
Start bar facing cut = 0.015 inches
Therefore, 13.047 inches/piece

Using Precut Standard "Slugs"
264 inches/14 inches per slug = 18.85, or 18 slugs per 22 foot bar
264 inches − (13.047 inches × 18) = 264 inches − 234.846 inches
= 29.154 inches loss per 22 foot bar including 0.953 inch chucking stock per slug
29.154/12 = 2.429 foot loss per bar.
At $7.49/foot this is an $18.19 stock loss per 22 foot bar including the chucking stock for each slug.

Using a Bar Fed Lathe and Process Cut Off
264 inches/13.047 inches = 20.23, or 20 parts per 22 foot bar 20 × 13.047 inches = 260.94 inches
264 inches − 260.94 = 3.06 inches chucking stock per bar 3.06 inches/12 = 0.255 foot per 22 foot bar.
At $7.49/foot this is a $1.91 stock loss per 22 foot bar

Therefore, a 22 foot bar would produce 18 shafts from precut standard material slugs or 20 shafts if the operation is performed on a bar feed machine. If the standard 14 inch material slug is used, there would be a 29.154 inch drop-off from the bar. This would be a part of the chip loss of the operation and its cost would be distributed to the 18 parts at 1.619 inches per part and be included in the purchase price of the 14 inch slug and the 22 foot bar.

If the operation is performed on a bar feed machine, the 20 pieces made from the 22 foot bar would produce a drop-off of 3.06 inches and two additional parts. The total material "nest" saving would be

(29.154 inches − 3.06 inches) = 26.094 inches, or $16.29 per bar

This will result in the production of two additional 13 inch worm wheel shaft parts per bar. This material saving is a "nest and gain" factor (F_{ng}) of 9.88% for this part based on the use of the bar feed manufacturing method instead of the standard design slug material allowance.

In addition, an allowance for quality problems and scrap must be included in the materials requirements planning function. There will still be some mistakes and scrap parts. This is a less controllable factor in the foundry where molds can fail or gas holes and chills can ruin some castings. Labor intensive operations and human controlled machinery also generate more errors than state-of-the-art computer controlled systems.

The magnitude of the scrap and error allowance (M_s) will depend upon the manufacturing methods used, the quality of the machinery and tooling, the skills and attitudes of the work force, the effectiveness of the quality management program, and the validity of the scrap projection procedure.

In this example operation, we have assumed that the materials from spoiled work can be recycled into the casting process. This means that the impact of errors on foundry material supplies will be to provide a small amount of recycled excess materials in the scrap iron and bronze supply. In general, this will have a very small impact on raw materials requirements or costs. It will have a tendency to be a safety factor. However, if it is a significant factor, the raw material purchasing requirements can be reduced to balance the inventory.

Conversely, if production schedules are to be met, there must be enough material and castings in the machine shop store room to make replacement worm shafts, worm wheel shafts, spacers, worm wheels, and housings if some are spoiled. The scrapped castings and bar and tubing material can be recycled into the foundry and used to make new castings to replace bad ones or those spoiled in machining. However, a scrap allowance must be included when calculating the bar and tubing material requirements for the machine shop.

In an earlier chapter, a scrap formula ($M_s = D\sqrt{n}$) was offered. This formula, which includes the square root of the lot quantity (n) multiplied by a difficulty factor (D), was offered to predict scrap losses in terms of required extra units in a production lot. This is a reasonably good approach when conventional machinery is in use (as in the foundry) and/or in labor intensive situations. However, the use of automated and computer controlled machinery significantly reduces the error factor in manufacturing and in most cases, very small scrap allowances are required. The shrinkage factor can also be based on experience and be expressed in percentage form (S%).

In consideration of these factors, one can apply the following formula to define the cost of the raw materials.

$$M_{sd} = [1 \pm F_{ng}][M_D \times (1 + M_s/n) \times (1 + S\%)]$$

where:

M_D = standard design material requirement in units of purchase (i.e., per pound, inch, foot, or each)
M_{sd} = adjusted material requirement per unit
M_s = scrap allowance in units per lot
n = number of units per lot (each)
S% = shrinkage factor in percentage
F_{ng} = "nest and gain" factor in percentage

Sample Product Cost Calculations

therefore

$$M_{sd} \times P \times N = \text{material cost per casting or part}$$

where

P = price per unit of purchase (i.e., per pound, inch, foot, or each)
N = units of material purchase per part (i.e., pounds, inches, feet, or each)

Scrap Iron:

$$M_{sd} \times \$0.50 \times N \text{ pounds per casting} = \$/\text{casting}$$

where

Purchase price per pound of scrap iron = P = $0.50
Standard material requirement per part or casting = M_D
Shrinkage allowance = $S\%$ = 0.0%
"Nest and gain" allowance = F_{ng} = 0.0%
Scrap allowance = M_s/n = 0% because of recycling

Bronze Ingot:

$$M_{sd} \times \$2.00 \times N \text{ pounds per casting} = \$/\text{casting}$$

Purchase price per bronze ingot pound = P = $2.00
Standard material requirement per part or casting = M_D
Shrinkage allowance in percentage = $S\%$ = 0.0%
"Nest and gain" allowance in percentage = F_{ng} = 0.0%
Scrap allowance in percentage = M_s/n = 0% because of recycling

Worm Shaft Bar:

$M_{sd} = (1 \pm F_{ng}) [M_D \times (1 + M_s/n) \times (1 + S\%)]$
$M_{sd} = (1 - 0.0569)[(16)(1 + 0\%)(1 + 0.00\%)]$
$M_{sd} = 15.0896$ inches per worm shaft @ 490 lb/cu foot.
Bar diameter = 3 inches; bar length (M_{sd}) = 15.09 inches
P = $0.70 per pound or $16.80 per foot
$S\%$ = 0% shrinkage allowance
F_{ng} = −5.69% "nest and gain" allowance if the part is to be made on the automatic bar lathe
M_s = an estimated scrap allowance of 0%

Worm shaft precut slug (16 inches) cost = $22.40
264 inches (22 feet) bar/16 inches = 16.5 slugs
16.5 × $22.40 = $369.60/16 = $23.10/shaft

or

$15.09/12 = 1.257$ foot @ 24 pound/foot $= 30.17$ pound $\times$ \$0.70
$= \$21.12$/shaft

This \$1.98 or 8.5% difference shows the effect of different methods and rounding at each step in the calculations. This is a critical factor in large volume operations. Therefore, a standard computation procedure should be defined to achieve a predictable and uniform error in the materials standards.

Worm Wheel Shaft Bar:

$M_{sd} = (1 \pm F_{ng})[M_D \times (1 + M_s/n) \times (1 + S\%)]$
$M_{sd} = (1 - 0.0988)[(14)(1 + 0\%)(1 + 0.00\%)]$
$M_{sd} = 12.62$ inches per worm wheel shaft @ 490 pound/cubic feet
Bar diameter $= 2$ inches: bar length $(M_{sd}) = 12.62$ inches
39.65 cubic inches/1728 $= 0.023$ cubic feet $\times$ 490 pounds/cubic foot
   $= 11.27$ pounds/shaft
11.27 pound/shaft $\times$ \$0.70 $= \$7.89$ materials/shaft
At 20 pieces per bar versus the standard material cost of a 14 inch slug
   @ 18 pieces per bar @ \$9.15 each
$P = \$0.70$ per pound or \$7.49/foot
$S\% = 0\%$ shrinkage allowance
$F_{ng} = -9.88\%$ "nest and gain" allowance if the part is to be made on the automatic bar lathe.
$M_s =$ an estimated scrap allowance of 0%
$M_{sd} = 11.27$ pound $\times$ \$0.70 $= \$7.889$ per piece

If a significant scrap rate exists in the machine shop, the scrap steel, 1045 chips, and bronze returned to the foundry can be treated as a material cost reducing variance. However, assuming good quality management, the scrap rate and material recycling impact should be insignificant.

These same calculations would be used to compute the material requirements for the worm wheel spacers which would be made from heavy wall steel tubing on the automatic lathe. For book space conservation we will assume the finished cost of these spacers, including materials, will be three dollars (\$3.00) each.

The computation of the parts' manufacturing costs will be discussed below.

E. COSTING PURCHASED PARTS AND COMPONENTS

Establishment of the cost of purchased parts appears to be a simple matter of soliciting quotations from suppliers. However, in order to achieve a precise cost for each item, it is necessary to compute the landed cost in the receiving department or supply storeroom. It is also necessary to calculate the cost impact of owning the inventory versus a Just-in-Time supply or "around the corner" availability from a supplier.

These comments on purchased materials and parts also apply to the acquisition of contingent shop materials (M_{ic}), and such burden materials (M_B) as shop supplies, maintenance and sanitary supplies, office stationery, tools, and to some extent, raw materials (M_D). In each case, the utilization cost of the item is more than its price or landed cost in the manufacturing plant storeroom because of the storage cost accrued by its residence in inventory.

Items which are to become a part of the product are usually classed in marketing as OEM (original equipment manufacturer) products or raw materials. The OEM products are sometimes distributed to other consumers as well, but that usually occurs through different marketing channels and with different price structures.

There are several marketing channels for the sale and distribution of OEM products. They are also often sold through multiple channels. The OEM supply system is very complex and the choice of a supplier or channel can significantly affect the landed cost of the product. The same marketing complexity also applies to many of the general supply items amd raw materials used by manufacturers.

As an example, steel bar stock of the type discussed above, can be purchased directly from the steel mill or through a steel warehouse or wholesaler. It can also be purchased in truckload quantities, in lots of a few bars on an order, or in slugs precut for production. In some cases, the steel mill will offer all levels of purchase. However, in most cases the small and semi-processed orders must be bought through a wholesaler.

In most cases, the prices will vary with the volume purchased, the source of supply, the mode of delivery, and the distance from the source to the user. These factors, and the cost of ordering, will affect the landed cost of the material at the receiving dock.

Another factor in material pricing is the frequency of ordering and the durability of the supplier/customer relationship. In the case of such items as bearings, fasteners, electrical fittings, valves, or steel bar, it is not uncommon for a user to arrange an annual contract with scheduled prices based on the sizes of the shipments and the total of the annual purchases. This type of supply arrangement is generally called contract purchasing. It helps to stabilize the cost of materials and often assures maintenance of both quality standards and availability, often on a Just-in-Time basis.

In the case of such components as bearings, fasteners, pipe fittings, electrical items, valves, motors, etc., there are also a variety of channels of supply. If a manufacturer specifies a particular brand of component for his product, it is often possible to negotiate a contract purchasing agreement with the manufacturer on an annual or long-term basis. This assures quality and availability, and can also help with cash flow.

In some cases, these contracts include consignment arrangements whereby the user holds supplier inventory and only pays for the items as they are withdrawn from stock into production. In such a case, the cost of warehousing is borne by the user while the vendor carries the cost of the inventory investment.

These long-term arrangements usually result in the least landed cost to the using manufacturer and they have the distinct advantage of assured supply and good quality control. They sometimes have the disadvantage of a fixed period cost in the face of more competitive pricing. Such situations are usually negotiable.

Some items and supplies are not sold directly by the manufacturer. Many items are only sold through an area wholesaler or distributor and cannot be purchased directly from the manufacturer. Other items can be bought from manufacturers' representatives or brokers who are independent agents of the manufacturer. In each of these marketing channels, the cost of the product is usually higher than a direct source purchase because of the middleman's commission. These channels can sometimes offer long-term or contract purchasing arrangements. However, the distribution channels are dictated by the suppliers' marketing policies.

Another source of supply is the specialized wholesaler or distributor. For example, there are companies who sell only bearings, but represent almost all of the bearing manufacturers in a particular sales territory. Other firms specialize in electrical equipment, hosing, power transmission equipment, instruments, plastics, etc. The advantage of dealing with such vendors lies in their expertise and knowledge of the sources of such items. They can, and often do, provide professional consultation in the design of a product which requires the application of the items which they sell. They will often guarantee availability and price for a period in return for a close and continuing vendor relationship.

As can be seen from these comments, the definition of the cost of a purchased component or part is specific to the item under consideration. In each case, the choice of vendor, the negotiation of prices and terms in the purchasing arrangement, the assurance of quality and availability of supply, and the reliability of the supplier are key issues. All of these factors have an impact on the landed cost of the material or component. The cost engineer must compute the cost of owning the inventory versus the availability and reliability of Just-in-Time supplies, compare the competitors and their prices, and define the clerical and handling costs involved in the procurement. Each item on the procurement list must be treated individually. Price is not the sole criterion.

In the development of the demonstration cost of the speed reducer, we will estimate or use supplier quoted prices for the purchased items. If we were to compute the true landed cost of the product, we would assume the cost of

Sample Product Cost Calculations

administering a purchase at $35 per order. However, in this case we will omit this cost element and assume a Just-in-Time supply situation with no in-house storage cost. We will also assume a contract purchasing arrangement with daily delivery of high-value items and a maximum inventory of two weeks of supply for such commodity items as bolts.

In the demonstration calculation, the landed costs of the purchased items for the speed reducer will be:

Purchased Parts

Two ball bearings	04-013-0013 @ $23.50
Two roller bearings	04-013-0014 @ $13.00
Two bearing seals	04-014-0015 @ $ 2.00
16 hex head bolts	04-015-0016 @ $ 0.60
Two allen socket plugs	04-015-0017 @ $ 0.40
Three ¼ inch square keys	04-016-0018 @ $ 0.50

F. COMPUTING LABOR RATES AND COSTS

As stated earlier, the cost of labor in the manufacture of a product should be based on the measurement of the labor time actually devoted to the process. However, in modern manufacturing, the worker often operates a number of machines and his time is distributed among them. The use of work measurement, work sampling, and such predetermined time systems as MTM or MOST, can help one to allocate the proper amount of labor time to each production operation. The time allocation will normally be in minutes or fractions of minutes per unit of production. The task of the cost engineer is to define the cost of a minute of labor so that only that labor which actually contributes to the manufacturing process or its operational support is charged to the cost or value of the product.

In defining the cost of a minute of labor, one must start with the employee's base wage or salary. This is usually defined as an hourly wage rate. For the purpose of this demonstration case, we will use a base labor rate of $20.00 per hour for skilled employees and a wage of $10.00 per hour for unskilled labor.

In addition, in the United States, there are a series of government imposed labor costs which must be added. These include:

Social Security taxes: The employer's share is currently 7.65% of wages and salaries. The employee has the same amount deducted from his wages. The cut off, before the latest tax law, was at annual earnings of $55,000.

Worker's compensation tax: This tax varies with the occupation. In this demonstration case, using Ohio laws, the machine shop rate would be

4.88% of wages and the foundry would be 13.18%. For this case we will assume all employees to be in the machine shop and use the 4.88% rate.

Unemployment compensation: There are two separate taxes in this cost. The federal unemployment tax is 0.08% of wages and in the case of Ohio, the tax is 3.1% of wages. Thus the total cost of this tax in Ohio is 3.18% of employee wages.

These taxes add 15.71% to the basic wages of the plant workers and salaried employees. In this case, the skilled workers would cost $23.142 per hour and the unskilled labor would cost $11.571 per hour. But that is not the whole cost of labor.

In most companies (and in all likelihood, the federal government will soon require that), management pays all or part of a health and hospitalization insurance program for employees. In some cases they also pay life insurance premiums. The average cost for large firms is in the range of $200 to $250 per month, per employee. In small firms it can be in the $500 range. For the purposes of this demonstration case, we will use a premium of $300 per month for health and life insurance. On the basis of a 2080 hour work year, this adds $1.73 per hour to each employee's wages. This makes the skilled wage $24.872 per hour and the unskilled wage rate $13.301 per hour.

If we add the cost of six holidays and two weeks of vacation, and use a 1952 hour work year, the 2080 hours of pay really costs another 6.15%. This makes the skilled wage $26.10 per hour or 130.5% of base pay and the unskilled pay $14.12 per hour or 141% of base pay.

Some companies also finance education, training, sick leave, child care, athletics, and social facilities. Military leave is also a legal requirement. The recent twelve week unpaid leave law has also added new costs for continuing fringes and paying replacement staff. For now, we will ignore these additional fringe costs.

It has been estimated that the average fringe benefit cost in the United States is about 35% of the base wage rate. This is probably low in many firms. However, we will use this figure in the demonstration case and assume that 35% is correct.

Therefore, the basic cost of a minute of work is the base wage rate plus 35%, divided by 60. In this case, that is $0.45 per minute for skilled labor and $0.225 per minute for unskilled workers. But there are other labor costs to consider.

Earlier discussion addressed the "for nothing" cost of the overtime premium, holiday work bonuses, night shift bonuses, and union steward pay. These "for nothing" costs are accrued by the workers in the course of the manufacturing operation. However, in this precision cost system, we will treat them as

Sample Product Cost Calculations

part of the overhead of the manufacturing business. These costs will be distributed to the product cost through the "use rate" based on the time use of the facility. These costs will only be charged directly to a product when the customer causes the use of overtime or night shift operations.

If a company chooses to regularly schedule more than forty hours per week, the "for nothing" charges for the overtime premium will be divided by the scheduled hours per work week and distributed over the whole week to establish a new standard hourly labor rate which will include the "for nothing" costs.

In summary, as discussed in Chapter 6, the cost of a minute of labor will include the usual fringe benefit charges and the "for nothing" costs will normally be charged to the product through the "use rate." The exceptions will be when the "for nothing" costs are customer driven or caused by regular shedules of more than forty hours per week.

G. COMPUTING WORK STATION "USE RATES"

All of the machining of the castings and parts will be performed in the manufacturing facility's machine shop. Computation of the cost of machining will be based upon the time use of the work station and its equipment and labor, plus contingent materials. This time use of the facility, or standard "rental" or "use rate" cost of the work station, will be charged to each part making or assembly operation in the manufacturing process in proportion to the time required to perform the specific operation. The costs of the castings and the production materials will be added to compute the part or product cost. The same time use of facility cost computation will apply to the foundry and assembly operations.

Computation of a work station's "use rate" begins with calculation of the basic space charge or "use rate" for the net operating space in the manufacturing plant. Computation of a space "use rate" for the net operating space of the machine shop will be the first step in the development of the parts' manufacturing costs.

For the purposes of the sample calculations in this chapter, let us assume a net operating space factor of 80% of the total area of the manufacturing building, not including the foundry. Let us also assume a building cost of $50 per square foot for a 100,000 square foot building. This will result in a building cost of $5,000,000 and a net operating space of 80,000 square feet. This will result in a constructive cost or value of $62.50 per square foot of the net operating space.

Assuming a straight line deprecation rate of 4% (25 years) and an annual working hour schedule of 2000 hours (40 hours per week times 52 weeks, minus 80 hours of vacation shut down), the basic hourly ownership cost or "use rate" for the net operating space is $0.00125 per square foot, per hour.

$$\frac{100,000 \text{ square feet} \times \$50}{80,000} = \$62.50$$

$62.50/25/2000 = \$0.00125$ per square feet, per hour

The general building burden or overhead and period expenses include such period costs as power, fuel, taxes, insurance, maintenance, and security. This cost will be estimated at $50,000 per annum. These burden expenses will be computed for the net operating space by dividing this figure by 80,000 square feet to get $0.625 per square foot, per annum or $0.0003125 per square foot per hour. By adding the $0.00125 for the building, this develops a basic space "use rate" of $0.00156 per square foot per hour for the "rental" of the net operating area of the manufacturing plant.

This space "use rate" is the "bucket" into which all of the other indirect costs of the work station are added or accrued. It is the core of the work station's "use rate" which is to be applied to the product.

This basic space "use rate," when multiplied by the area of the work station, develops the space charge or "rent" for the work station. Thus, if a work station occupies 100 square feet, its space "use rate" will be $0.156 per hour.

The manufacturing burden rate (supervision, inspection, setup, materials handling, maintenance, clerical, etc.), and the cost of general and administrative expense must also be added in order to develop the work station's time use of facility "use rate."

If we assume a supervisory salary of $40,000 per annum with a fringe benefit rate of 35%, the hourly cost of supervision will be $27.00 based on a 2000 hour work year. With one person employed in inspection, and one in set-up, and paying the same salary rate for the inspectors and set-up men, the plant total for this indirect manufacturing overhead labor will be:

[($40,000 × 1.35)/2000] × 3 = $81.00 per hour

Using the same procedure for two materials handlers and three maintenance mechanics, and using pay rates of $20,000 and $30,000 respectively, we have hourly plant burden labor rates of:

[($20,000 × 1.35)/2000] × 2 = $27.00 per hour
[($30,000 × 1.35)/2000] × 3 = $60.75 per hour

By dividing the total of these plant burden labor costs ($168.75 per hour) by the 80,000 square feet of net operating area, the result is an additional charge of $0.0021 per hour per square foot for factory burden or overhead labor ex-

Sample Product Cost Calculations

pense. This brings our basic space "use rate" to $0.00366 per hour per square foot.

Other burden costs such as contingent materials, manufacturing clerical staff, security and fire defense, factory personnel administration, "for nothing" costs, and factory management must also be included in the work station's "use rate" charges.

Let us assume that the additional burden expenses total $450,000 per annum for our factory. On the basis of the 80,000 square foot net operating space, the hourly cost of the burden items would be $5.625 per square foot, per annum or $0.00281 per square foot per hour. This results in a cost of $0.0000468 per minute. This brings the space "use rate" to $0.00647 per square foot per hour, without the machinery.

The "use rates" for the machinery and equipment used in the work station must be included in the work station's "use rate." In a real life situation, the cost of the equipment in a work station may vary from a few hundred dollars to over a million dollars. However, for our example, we will assume that the automated machining operations will require either a $300,000 or $250,000 equipment investment.

We will use the straight line method of depreciation and assume a ten year depreciation "life" for the equipment. If the equipment in the work station has a capital value of $300,000, the hourly "use rate" for the equipment would be $300,000/10/2000 or $15 per hour. This is $0.25 per minute. In the case of the $250,000 machine, the "use rate" would be $12.50 per hour or $0.208 per minute.

As discussed in earlier chapters, the recommended method for applying the general and administrative expense (G & A) to the product is based on the principle that G & A is the cost of managing the total investment in the enterprise. Therefore, the G & A is applied to the "use rate" on the basis of the G & A dollars per year, per dollar of capital investment in the facility. Investment is defined for this purpose as the dollars originally invested in the facility and not the depreciated value of the assets.

For the purpose of this example, let us assume that there are twenty work stations, each using 300 square feet of space and $300,000 worth of equipment, and another twenty work stations with 200 square feet and $250,000 in equipment. In addition, there will be ten assembly work stations with 100 square feet and $5,000 in equipment. This will require a total manufacturing area of 11,000 square feet for a capital investment of $550,000 in space and $11,050,000 in machinery and equipment. In addition, the plant will have 89,000 square feet of warehousing, offices, toilets, aisles, maintenance shops, and shipping and receiving space at $50 per square foot or $4,450,000 in warehouse and supporting building capital investment. The total capital investment will be $16,050,000.

The G & A annual expense for the executive staff and their support is assumed to be $2,000,000 per annum. Dividing the $2,000,000 by the $16,050,000 investment will give us a G & A factor of $0.125 per dollar per year, or $0.0000623 per dollar per hour. This general and administrative expense must be added to the other costs to produce the complete work station facility "use rate."

For example, if the space investment in work station A is 300 square feet of building space times $50 per square foot, or $15,000, and the investment in the work station's equipment is $300,000, this makes the total capital investment in work station A equal to $315,000. The basic space and equipment "use rate" is $15.30 per hour or $0.255 per minute.

For work station A, this will produce a G & A charge of [(300 square feet × $50) + $300,000] × $0.0000623 = $19.6245 per hour or $0.327 per minute added to the work station's basic $15.30 per hour or $0.255 per minute "use rate." This results in a total work station "use rate" of $0.582 per minute, not including labor or burden expenses.

Similarly, if the investment in work station B is based on 200 square feet of space at $50 per square foot or $10,000, plus $250,000 in machinery and equipment for a total of $260,000 in capital investment, the G & A charge will be:

[(200 square feet × $50) + $250,000] × $0.0000623 = $16.198 per hour or $0.2699 per minute added to the work station's "use rate."

Using the preceding sample figures, let us describe two typical work stations.

1. If the work station utilizes 300 square feet of space and a $300,000 machine, the work station's "use rate" would be calculated as follows:

Space @ 0.00125/sq ft/hr × 300 sq ft	= 0.375/hr
Building exp @ 0.0003125/sq ft/hr × 300	= 0.09375/hr
Burden labor @ 0.0021/sq ft/hr × 300 sq ft	= 0.63/hr
Factory burden exp @ 0.00281/sq ft/hr × 300	= 0.843/hr
Equipment "use rate"	= 15.00/hr
General & administrative expense	= 19.6245/hr

   The "use rate" for work station A would then total to be $36.56625 per hour, or $0.609 per minute.

2. If the work station utilizes 200 square feet of space and a $250,000 machine, the work station's "use rate" would be calculated as follows:

Space @ 0.00125/sq ft/hr × 200 sq ft	= 0.25/hr
Building exp @ 0.0003125/sq ft/hr × 200	= 0.0625/hr

Sample Product Cost Calculations

Burden labor @ 0.0021/sq ft/hr × 200 sq ft = 0.42/hr
Factory burden exp @ 0.00281/sq ft/hr × 200 = 0.562/hr
Equipment "use rate" = 12.50/hr
General & administrative expense = 16.198/hr

The "use rate" for work station B would then total to be $29.9925 per hour, or $0.4998 per minute.

The direct manufacturing labor cost must be added to this total work station "use rate" to produce the complete time use of facility cost of each minute used in the operation.

We will compute some of the manufacturing costs as a part of this demonstration chapter. However, some of the machining costs will be assumed in order to limit the length of the chapter.

H. PROCESS DEFINITION FOR "MAKE" PARTS

In order to demonstrate the precision cost system, we will "manufacture" and cost four of the parts of the worm gear speed reducer. We will assume an in-house iron foundry and bronze foundry.

In the manufacturing process we will use $300,000 CNC, multiple axis, multiple tool, machining centers (Figure 16.2) to process the gear reducer's main housing casting and other iron castings. We will use $250,000 CNC, bar fed, turn/mill lathes to process the worm wheel shaft and worm shaft. The CNC lathe used to finish the worm wheel casting into a gear blank will cost $200,000. We will use $100,000 CNC gear hobbing machines in 200 square foot work stations to cut the worm wheel teeth. The cost of broaching the worm wheel's key way will not be computed. It will be assumed to be $4.00 per part.

As stated earlier, there will be many of each of these work stations and the production volume passing through the factory will not impact our cost computations. In order to limit the length of this chapter, we will assume a finished part cost for all other components of the speed reducer. In a later section, we will address assembly operations.

The iron gear housing castings (Part No. 05-017-0019) will be delivered to the machine shop from the foundry by forklift truck in four foot square "common denominator" pallet tubs containing 20 castings per tub. Each casting will weigh 20 pounds. Its delivered cost in the machine shop is $14.00 per casting.

Two tubs of castings will be located on a turntable at each machining center, and the empty tub will be removed and replaced while the other tub's contents are being processed. The finished castings will also be placed on a pallet on a second turntable and the full pallets will be removed while the second pallet is being loaded. There will be no machine downtime for casting tub place-

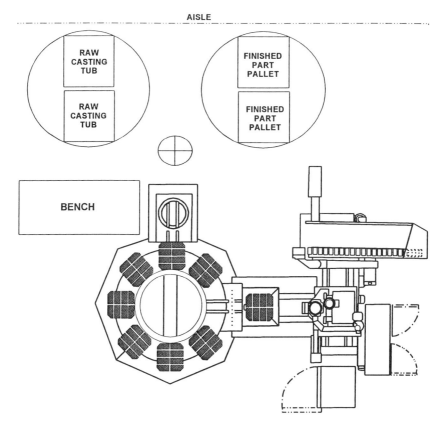

Figure 16.2 Milling center general arrangement.

ment or finished part pallet removal. This turntable delivery and removal technique will be used at all of the machining work stations.

The selection of rough castings from the tubs and the placement of finished parts on the pallets will be accomplished during the machining cycle and will not be an addition to the load/unload cycle time. The finished castings will be inspected by the operator during the machining cycle and before placement on the pallet.

The machining center (Figure 16.2) will be equipped with a fifty position tool changer, tooling for the required operations, an indexing table fixture, and an eight position, turntable type, automatic load/unload station. The machine will be tooled and programmed to perform all of the operations on the castings.

Sample Product Cost Calculations

The machining cycle, after the first piece, will be as follows:

The castings will be machined in two steps. The bottom and the shaft holes, and their related drilled and tapped holes, will be machined first on the top half of the tool pallet's two operation single fixture. The semi-finished casting will then be removed and inverted, and transferred to the bottom half of the tool pallet fixture for top finishing. The operator will remove both the semi-finished (first operation) part and the finished part, reload the semi-finished casting for the second operation, and load a raw casting for the first operation.

While the machine is performing the two operations, the operator will inspect the finished part and place it on the pallet, and retrieve a raw casting from the tub and mount it on the next tool pallet station.

While the machine is processing the castings located in the two work positions on the machine's tool pallet, the following loading operations will be performed at the loading turntable.

Work Piece Load/Unload Cycle

1.	Remove finished and semi-finished parts from machine's work position to the turn-table	0.20 min
2.	Rotate turntable and insert casting and semi-finished part into machine	0.30 min
3.	Place finished part on bench	0.10 min
4.	Take semi-finished part off turntable fixture	0.15 min
5.	Reload semi-finished part to turntable fixture	0.75 min
6.	Retrieve rough casting from bench or tub	0.10 min
7.	Load rough casting to turntable fixture	0.75 min
	Work piece load/unload cycle time	2.35 min

Operator Parallel Work Cycle

8.	Visually inspect finished casting	0.50 min
9.	Place finished casting on pallet	0.30 min
10.	Retrieve rough casting from tub	0.30 min
	Operator parallel work cycle time	1.10 min
	Operator internal work cycle total	3.45 min
	Operator idle time during machining	6.30 min
	Operator time	*9.75 min*

Machine Operation Work Cycle

11.	Turntable rotates and loads machine	0.30 min

12. Machining run or cycle time (bottom)

Finish bottom of casting	1.10 min
Bore worm shaft bearing holes	1.10 min
Face bearing hole cover seats	1.05 min
Bore worm wheel bearing holes	1.10 min
Face worm wheel hole cover seats	1.10 min
Drill and tap bolt and mount holes	1.80 min
Swap finished parts and castings and change program from raw casting to semi-finished casting	0.50 min
Machining run time (bottom)	*8.05 min*

13. Machining run or cycle time (top)

Finishing top of casting run time	1.20 min
Drill and tap two plug holes	0.50 min
Machining run time (top)	*1.70 min*

TOTAL MACHINE CYCLE TIME FOR TWO HOUSING CASTINGS 9.75 min

The materials handling of the tubs and pallets is internal to the machining cycle at all machining work stations. The loading and unloading of the turntable fixtures and machine loader is internal to the machining run time. Therefore, the work center "use rate" and labor cost must be applied to the machine run or cycle time to calculate the part cost. This computation will be shown in the next section of this chapter.

The worm shaft (Part No. 03-003-0011) will be made on an automatic CNC turn/mill lathe with a 3 1/2 inch spindle hole and an automatic bar feed capable of handling six, 22 foot long bars of 2 or 3 inch diameter, 1045 CRS. The lathe's turret will have ten tool positions including five that are capable of mounting rotating tools such as milling cutters, drills, and taps. The lathe will have a powered sliding tail stock with a retractable rotating center. The cross slide will have four tool holding positions and the lathe will have an automatic part removal robot arm.

The bar feed system will be loaded from the aisle each time the machine starts working on the last bar in the bundle. It will be loaded with a specially modified, four way, reach forklift truck and the lathe operator will remove the straps from the bundle. The three bar (3 inch diameter) bundles will weigh 1,584 pounds. Each three inch, 22 foot bar will weigh 528 pounds.

The finished parts will be placed in specially vacuum formed and reusable plastic trays and stacked on pallets. The pallets will be located on a two position turntable. The full pallets will be removed by a forklift truck and replaced with an empty pallet and a pile of plastic trays while the second pallet is being loaded with finished parts.

Sample Product Cost Calculations

The machining cycle for the one piece worm and worm shaft, after the first piece, will be as follows:

Machining Cycle Time

1. The bar will be positioned against the stop by the machine's bar feed program and the machining cycle will start.
2. Face bar end . . . . 0.20 min
3. Drill center hole and position center . . . . 0.70 min
4. Turn shaft ends to worm pitch diameter . . . . 1.60 min
5. Turn worm thread form to finish size . . . . 2.50 min
6. Contour and finish shaft end diameters . . . . 1.70 min
8. Chamfer shaft ends . . . . 0.30 min
9. Mill ¼ in. × 2 in key way . . . . 0.60 min
10. Cut off and robot remove part . . . . 0.60 min
11. Feed bar for next part . . . . 0.20 min
    TOTAL MACHINING CYCLE TIME FOR PART   8.40 min

Operator Cycle Time

12. Operator removes part from robot drop during bar feed cycle . . . . 0.20 min
13. Operator cuts bar bundle strapping during machine work cycle and disposes of straps. . . . . 0.80 min
14. Operator visually inspects part . . . . 0.50 min
15. Operator places part in plastic tray and loads pallet during machine cycle. . . . . 0.20 min
    Operator time for part   *1.70 min*
16. Operator idle time during machine cycle . . . . 6.70 min
    OPERATOR INTERNAL WORK CYCLE TOTAL   8.40 min

On the basis of this work cycle, it appears that the operator can manage two machines and still have some idle time. The total operator time for two machines would be 3.4 minutes per cycle and the cycle time is 8.40 minutes. This would leave 5.0 minutes of idle operator time per cycle. The machines could be placed in facing positions with the operator between them and aisle access behind the machines for bar loading. The pallet turntable could be placed at one end of the work cell with aisle access. The two machine cells could easily fit in a 400 square foot work station.

The worm wheel shaft (Part No. 03-010-0010) will be made on an identical CNC turn/mill lathe. The bar feed system will also be loaded from the aisle each time the machine starts working on the last bar in the bundle. It will be loaded with the same specially modified, four way, reach forklift truck and the lathe operator will remove the straps from the bundle. The six bar (2 inch diameter) bundles will weigh 1,412.4 pounds. Each 2 inch, 22 foot bar will weigh 235.4 pounds.

These finished parts will be placed in specially formed reusable plastic trays and stacked on pallets. The pallets will be located on a two position turntable. The full pallets will be removed by a forklift truck and replaced with an empty pallet and a pile of plastic trays while the second pallet is being loaded.

The machining cycle for the worm wheel shaft, after the first piece, will be as follows:

MACHINING CYCLE TIME

1. The bar will be positioned against the stop by the machine's program and the machining cycle will start
2. Face bar end   0.15 min
3. Drill center hole and position center   0.70 min
4. Finish turn shaft to design diameter   1.60 min
5. Chamfer shaft ends   0.30 min
6. Mill three ¼ inch by two inch keyways   1.70 min
7. Cut off and robot remove part   0.60 min
8. Feed bar for next part   0.20 min
   Machine operation time cycle for worm wheel shaft   *5.25 min*

OPERATOR CYCLE TIME

9. Operator removes part from robot drop during bar feed cycle   0.20 min
10. Operator cuts bar bundle strapping during machine work cycle and disposes of straps   0.80 min
11. Operator visually inspects part   0.50 min
12. Operator places part in plastic tray and loads pallet during machine cycle   0.20 min
    Operator cycle time   *1.70 min*
13. Operator idle time during machine cycle   3.55 min
    Operator Cycle Time for Part   5.25 min

On the basis of this work cycle, it appears that the operator can manage two machines and still have some idle time. The total operator time for two machines would be 3.4 minutes per cycle with the cycle time as 5.26 minutes. This would leave 1.85 minutes of idle operator time per cycle. The machines could be placed in facing positions with the operator between them and aisle access behind the machines for bar loading. The pallet turntable could be placed at one end of the work cell with aisle access. The two machine cell could easily fit in a 400 square foot work station.

The bronze worm wheel casting (Part No. 05-018-0023) will come to the machine shop from the company foundry at a cost of $95.00 each. They will

Sample Product Cost Calculations

be in a 4 foot square "common denominator" shop tub with 20 gear castings per tub at 44 lb each (880 pounds per tub). The worm wheels (gears) will be machined on the same type and size CNC turn/mill lathe as the shafts. However, the machine will be equipped with a soft jaw chuck machined to hold the casting on both sides and a part reversing and removal robot. There will be no requirement to use the bar feed system on these chucking machines.

The machining cycle for the worm wheel (gear) casting, after the first piece, will be as follows:

MACHINING CYCLE TIME

The 44 pound worm wheel casting will be manually positioned into the hydraulically actuated and preshaped soft jaw chuck which will grip the outer diameter of the worm wheel. The machine will be equipped with a positioning arm which will be used to hold and align the casting during the chucking operation.

1. Chuck worm wheel casting — 1.50 min
2. Face hub end — 0.15 min
3. Drill one inch center shaft lead hole — 0.70 min
4. Bore two inch diameter shaft hole — 1.10 min
5. Finish turn hub outside diameter — 0.50 min
6. Finish turn and countour gear face — 1.20 min
7. Chamfer hub hole and hub O.D. — 0.20 min
8. Robot pick-up and reversal of piece by using an expanding arbor type end effector in the two inch shaft hole. — 0.20 min
9. Place reversed semi-finished casting on turret mounted expanding centering arbor. — 0.30 min
10. Turret moves semi-finished casting to chucking position and chuck grips finished part of hub's outside diameter — 0.50 min
11. Finish turn and contour gear face — 1.20 min
12. Finish turn gear outside diameter — 0.70 min
13. Chamfer gear outside diameter — 0.20 min
14. Robot pick up and removal of finished piece to holding tray and machine stops for reloading. — 0.50 min

   Machining cycle time — *8.95 min*

Operator Cycle Time

15. Operator retrieves casting from tub and loads into chuck — 0.80 min
16. Operator removes finished part from robot's holding tray drop during machining cycle — 0.50 min
17. Operator visually inspects part — 0.50 min

18. Operator places part in plastic tray and loads pallet
during machine cycle   0.30 min
    Operator cycle time   *2.10 min*
19. Operator idle time during machine cycle   6.85 min
    TOTAL OPERATOR CYCLE TIME   8.95 min

On the basis of this work cycle, it appears that the operator can manage three machines and still have some idle time. The total operator time for three machines would be 6.30 minutes per cycle and the cycle time is 8.95 minutes. This would leave 2.65 minutes of idle operator time per cycle.

The machines could be placed in a triangular facing work cell arrangement with the operator in the center. The pallet turntable could be placed at an open end of the work cell with an aisle access. The three machine cell could easily fit in a 600 square foot work station.

Upon completion of the turning operations, the bronze worm wheel will weigh less than 40 pounds and will be moved to the hobbing machine in pallet lots of fifty gear blanks. The hobbing operation will take 8 minutes, and four machines will be operated by each worker. The hobbing "use rate" will be $0.40 per minute and the labor rate will be $0.45 per minute. The resulting hobbing operation cost will be $6.80 per worm wheel.

Although not detailed here, the machining of the other iron castings and the worm wheel spacers would follow similar procedures. The cost of machining these parts will be estimated to limit the length of this chapter.

I. COMPUTING MANUFACTURING OPERATIONS' COSTS

In order to compute the cost of each manufacturing operation, it is necessary to apply the work center's "use rate" and the direct labor cost to the actual or standard work cycle time. It is then required to add the cost of direct material and any specific supporting activities and/or contingent materials.

On the basis of the sample calculations presented in earlier chapters, we will assume the space "use rate" in this machine shop to include building ownership costs, burden, supervision, burden labor, contingent and burden materials, taxes, utilities, insurance, and all other period cost factors except general and administrative costs. Using an assumed 100,000 square foot building at $50.00 per square foot and a net operating space of 80,000 square feet, and using the sample calculations in Chapter 12 as a source, the basic space "use rate" will be $21.325 per square foot per annum or $0.012172 per square foot per hour based on a 1752 hour year. This is $0.000202 per square foot per minute.

The equipment "use rate" is computed on the basis of a ten year straight line depreciation rate and a 10% interest charge, plus 15% annual maintenance

Sample Product Cost Calculations

cost. Using $100,000 as a base, and 1752 hours per year, this generates a "use rate" of $19.92/hour, per $100,000 of equipment investment. This is $0.332/minute.

Referring back to Chapter 13, we will assume the general and administrative cost to be $0.00001 per invested dollar per hour. Using these figures, the total "use rate" for each of these work stations will be:

1. MILLING CENTER:

    Work station area − 200 sq ft @ $50.00/sq ft
    Milling center machine − $300,000 per unit

    Space
    200 sq ft @ $0.000202/min   = $0.0404/min
    G & A
    G & A @ $0.00001/$/hour
    $310.000 × 0.00001 = $3.10 per hour =$0.05116/min
    Equipment
    $300,000 × $0.332/min/$100,000   = $0.996/min
    MILLING CENTER "USE RATE"   = $1.08756/MIN

2. TURN/MILL BAR MACHINE:

    Work station area − 200 sq ft @ $50.00/sq ft
    Milling center machine − $250,000 per unit

    Space
    200 sq ft @ $0.000202/min   = $0.0404/min
    G & A
    G & A @ $0.00001/$/hour
    $260.000 × 0.00001 = $2.60 per hour =$0.04333/min
    Equipment
    $250,000 × $0.332/min/$100,000   = $0.830/min
    TURN/MILL CENTER "USE RATE"   = $0.91373/MIN

3. TURN/MILL CHUCKING MACHINE:

    Work station area − 200 sq ft @ $50.00/sq ft
    Turn/Mill Chucking Center Machine − $100,000 per unit

    Space
    200 sq ft @ $0.000202/min   = $0.0404/min
    G & A
    G & A @ $0.00001/$/hour
    $110.000 × 0.00001 = $1.10 per hour =$0.01833/min
    Equipment
    $100,000 × $0.332/min/$100,000   = $0.332/min
    TURN/MILL CHUCKING CENTER "USE RATE" = $0.39073/MIN

In the case of the parts described above, the cost of contingent materials, general materials handling, and inspection operations have already been included in the burden portion of the work station "use rates." However, in the manufacture of the shafts, an additional handling operation has been introduced. A special and specifically assigned bar handling machine is included in these operations. The cost of this materials handling operation must be computed and added into the cost of the manufacturing operation.

Assuming an equipment cost of $40,000 for the special bar stock handling four way reach truck, and using a ten year straight line depreciation rate, its basic annual cost will be $4,000. The cost of interest, insurance, power, taxes, and maintenance must be added to this. If we assume that these expenses are 15% of the cost of the vehicle, or $6,000 per year, the total equipment ownership cost will be $10,000 per annum.

If we consider the fork operator to be semi-skilled or unskilled, the applicable wage rate, including fringes, will be $13.50 per hour or $13.50 × 2080 = $28,080 per annum. Thus the annual cost of the bar handling machine will be $10,000 + $28,080 = $38,080. This will give a standard hourly cost of $19.51 based on a 1952 hour work year or $0.325 per minute. These machines support all of the bar feed lathes in the plant.

The bar bundle handling and feeder loading requires 0.50 minutes to pick up the bundle in the raw material store room, a loaded movement of 300 feet at 200 feet per minute (1.5 minutes), a bar feed tray loading time of 3 minutes, and an empty return trip of 300 feet (1.5 minutes). Therefore, the time required to load the bar feed tray will be 6.5 minutes at $0.362 per minute for a cost of $2.353 per bundle moved.

In the handling of the three bar bundles of three inch stock, there are 17 parts per bar or 51 parts per move. The six bar bundles of two inch bars make 19 parts per bar or 114 parts per bundle. Thus, the cost of the move is $0.046 per part for the worm shafts and $0.02 per part for the worm wheel shafts.

Let us assume that the operators of each of the CNC machines are classed as skilled workers and their wage rates, including fringes will be $27.00 per hour or $0.45 per minute.

Referring back to the previous sections, and applying the machining work station cycle times, work station "use rates," labor costs, and material costs, the cost of making each of these parts will be:

Gear box housing (2 pieces per cycle):
Material, iron casting $14.00/part
Machining cycle time, 9.75/2 min/piece
Work station "use rate", $1.08756/min × 9.75/2 min = $ 5.30/part
Labor wage cost, $0.45/min × 9.75/2 min   = $ 2.19/part
COST PER HOUSING   = $21.49/part

Sample Product Cost Calculations

Worm shaft:
Material, inch CRS 1045 rod	= $22.15
Machining cycle time, 8.40 min/piece	
Work station "use rate" $0.91373/min × 8.40 min	= $ 7.68/part
Labor wage cost, $0.45/min × 8.40 min	= $ 3.78/part
Bar feed handling	= $ 0.04/part
COST PER WORM SHAFT	= $33.65/part

Worm wheel shaft:
Material, 2 inch CRS 1045 rod	= $ 8.74
Machining cycle time, 5.26 min/piece	
Work station "use rate," $0.91373/min × 5.25 min	= $ 4.797/part
Labor wage cost, $0.45/min × 5.25 min	= $ 2.362/part
Bar feed handling	= $ 0.02/part
COST PER WORM WHEEL SHAFT	= $15.919/part

Worm wheel:
Material, bronze casting	= $95.00
Machining cycle time, 8.95 min/piece	
Work station "use rate," $0.39073/min × 8.95 min	= $ 3.497/part
Labor wage cost, $0.45/min × 8.95 min	= $ 4.027/part
Hobbing operation cost	
Labor, $0.45/min × 8.00 min	= $ 3.60/part
Hobbing "use rate," $0.40/min × 8.00 min	= $ 3.20/part
Broaching key way-estimated	= $ 4.00/part
COST OF WORM WHEEL	= $113.32/part

It is also feasible to compute the cost of each element of the machining operation. For example, if one were to compare the cost of turning the worm thread with the cost of rolling that shape, it would be necessary to isolate the cost of the thread cutting operation. In this example the thread cutting operation cost would be:

Turn worm thread form to finish size	= 2.50 min/piece
Work shaft station "use rate," $0.91373/min × 2.50 min	= $2.284/part
Labor wage cost, $0.45/min × 2.50 min	= $1.125/part
Cost of the thread turning operation	= $5.909/part

In costing the rolling operation, the cost of multiple handling of the semi-finished part between the turning machine and the rolling operation would be added to the cost of thread rolling in order to make the comparison.

These same computations will be applied at each step in the manufacturing process as the parts and components progress through the system. In each case, the cost of the individual operation will be added to the inventory valuation of the work-in-process part as it accumulates cost and progresses toward completion. The accumulated operations' costs will become the finished part cost in the WIP inventory. The finished part cost will be the material cost which is input to the assembly operation.

J. COMPUTING ASSEMBLY OPERATIONS' COSTS

Development of the cost of assembly operations will follow the same general procedure as has been used in the costing of the machining operations. The primary differences will be in the relationships between the size of the capital investment in the work station and the amount of direct labor used.

However, in many modern manufacturing systems, the assembly process is mechanized or automated. The capital investment in robots, programmable conveyors, special tooling, fixtures, automated welders, automated testers, and other devices can be substantial. For the purposes of this book, we will assume a semi-mechanized assembly process.

The numerical, indented manufacturing bill of materials for the speed reducer will be structured as follows:

Finished Product	
Worm gear speed reducer	01-001-0001
Major Sub-Assemblies	02-XXX-XXXX
Housing	02-002-0002
Housing	03-002-0005
Front Cover Plate	03-006-0006
Solid End Plate	03-006-0007
Seal Hole End Plate	03-006-0008
16 Hex head bolts	04-015-0016
Two allen socket plugs	04-015-0017
Worm Wheel Assembly	02-004-0004
Worm wheel shaft	03-010-0010
Bronze worm wheel	03-009-0009
Two worm wheel spacers	03-012-0012
Two roller bearings	04-013-0014
One bearing seal	04-014-0015
Two ¼ inch square keys	04-016-0018
Worm Shaft Assembly	02-003-0003
One piece worm shaft	03-003-0011
Two ball bearings	04-013-0013

Sample Product Cost Calculations

One bearing seal	04-014-0015
One ¼ inch square key	04-016-0018
Semi-Finished Parts	05-XXX-XXXX
Housing casting	05-017-0019
Front cover casting	05-017-0020
Two solid end castings	05-017-0021
Seal hole casting	05-017-0022
Bronze gear casting	05-018-0023
Raw Materials	06-XXX-XXXX
Scrap iron	06-024-0024
Bronze ingots	06-025-0025
1045 CRS steel 3 inch rod	06-026-0026
1045 CRS steel 2 inch rod	06-026-0027

We will use the worm wheel and shaft sub-assembly as the sample operation to demonstrate the computation of assembly costs. It will be assembled on a programmed pallet type conveyor equipped with appropriate assembly fixtures, parts' feeders, and robots. The operation can be described as follows:

The bronze worm wheel will arrive at the assembly line on a pallet containing fifty units in plastic trays of twenty-five pieces each. The pallet will be placed on a two pallet turn table by the forklift truck.

A robot will lift the worm wheel from the pallet tray, align the key way with the fixture, and place the gear in a nest type fixture which will clamp and center the gear on the twelve inch square conveyor pallet (1.05 minutes). The conveyor pallet will move to the next station (0.30 minutes). At the next station, a second robot will pick up a worm wheel shaft from its pallet tray, also located on a turntable, and place it in a key way orientation device. The robot will then insert the end of the shaft into the shaft hole of the worm wheel with the shaft key way aligned with the gear key way. The shaft will be inserted to a point where the end of the center shaft key way is 0.005 inches from the face of the worm wheel hub (1.20 minutes).

At that point, and while the shaft is being held in position, another robot will insert the 1/4 inch key into the shaft's slot and clamp it in place (0.50 minutes). The shaft handling robot will then push the shaft into position in the worm wheel (0.30 minutes). The conveyor pallet will move on to the next work station (0.30 minutes). At the next work station, a robot will grasp the worm wheel and shaft assembly and remove it from the nesting fixture. It will be rotated into a shaft horizontal position and mounted in a holding fixture on a conveyor pallet (0.20 minutes)

The conveyor pallet will move on to the next work station (0.30 minutes).

The next work station will be equipped with parts feeders which will deliver two shaft spacers and two tapered roller bearings to within easy reach of both hands of an assembler. The assembler will place the spacers on both ends of the shaft and slide and press the roller bearings on the shaft against the spacers. The operator will then gauge check the positions of the bearings (2.50 minutes). The conveyor pallet will move on to the next work station (0.30 minutes).

The worm wheel and shaft assembly will be picked up by a robot and placed in a molded plastic holding tray on a pallet. There will be twelve assemblies per tray and they will weigh approximately 60 pounds each. The same robot will also place an empty tray on top of the filled tray every 12 cycles to build a pallet load of 24 assemblies. The pallet will be on a turntable with an adjacent stack of trays (0.50 minutes).

A forklift truck will remove filled pallets and replenish the pallets and trays. The same forklift will deliver inbound parts to the assembly work stations. The forklift truck will have the same "use rate" cost as the unit supplying the bar stock to the lathes and it will be fully dedicated to serving two worm wheel and shaft assembly work stations at $0.317 per minute. It will spend approximately half of its time with each work station. This activity will be internal to the assembly cycle.

The dominant work cycle of the work station will be the manual assembly of the spacers and bearings. Each of the other operations is shorter than the manual assembly task. The pallets on the conveyor, and the inspector, will require some waiting time to achieve balance on the line.

It is assumed that the capital equipment described above, which includes a programmable assembly conveyor, four robots, several special fixtures, four parts feeders, three turn tables, and miscellaneous tools and furniture, will have an initial capital cost of $300,000. The area of the assembly work station will be 300 square feet. This space and equipment will generate a work station "use rate" of $0.61 per minute based on a calculation similar to those demonstrated above.

The single worker will be considered skilled and will have a wage cost of $27.00 per hour including fringes. This will be a labor cost of $0.45 per minute.

Therefore, the worm wheel and shaft assembly cost per cycle will be based on:

2.80 minutes × ($0.61 + $0.45) = $2.968

per sub-assembly plus half of the forklift services (1.40 minutes) per work station, per part. The forklift "use rate" is $0.317 per minute including labor. This adds a cost of $0.44 per part. However, the forklift operating time is internal to the assembly time.

Sample Product Cost Calculations

The total cost is therefore $3.41 per sub-assembly.

At 2.80 minutes per cycle, each work station should assemble 171 units per eight hour shift.

By using a similar analysis procedure, and using four mechanized work stations for the complete assembly, it is estimated that the cost of assembling the speed reducer will require a composite cycle time of approximately fifteen minutes per unit. Using the same "use rate" and labor cost standards, the complete assembly cost would be:

15 minutes × ($0.61 + $0.45) = $15.90 each

In addition, two forklift trucks will serve the four assembly work stations at $0.317 per minute for 7.5 minutes per unit. This supporting work will not add elapsed time to the process but it will add cost. It will add $2.38 in assembly handling cost per speed reducer.

This results in a fully absorbed total assembly cost of $18.28 per speed reducer.

K. COMPUTING WAREHOUSING/STORAGE COSTS

As stated earlier, there are three or four types or classes of storage operations in the manufacturing process. These usually include raw materials storage, receiving or supply stores, work-in-process storage, and finished goods storage. In each case the cost of operations includes the ownership or "use rate" cost of the cubic space occupied, the in and out handling cost, and the cost of order picking and documentation. The foregoing section on warehousing discussed the finished goods operations. In this demonstration of the cost system we will use the work-in-process parts storage activity as the basis for the computations. Raw materials and supply stores operations require similar calculations which are based on slightly different physical storage and handling characteristics.

In each of the above described manufacturing and assembly operations, we have assumed a pallet handling system. This and other "common denominator" handling modules (i.e., tote boxes, barrels, cartons, etc.) assure the use of a homogeneous materials handling system and versatile handling machinery.

In this case we have assumed the use of manned forklift trucks in the one ton capacity class. The use of automated guided vehicles would not change the cost computation procedure but it would increase the "use rate" for the machines and eliminate the labor element from the cost. An exception is a specialized machine as proposed for handling the bar stock into the lathes.

The nature of the WIP operation, which includes trays of heavy parts on pallets with low pallet load heights and limits on pallet stackability, will result in the need for a pallet rack storage system with short height openings. For uniformity, we will assume a pallet rack configuration which is based on a 48 inch by 48 inch, four way, pallet or shop tub which is loaded 24 inches high plus the 6 inch high pallet. This results in a unit load cube of 40 cubic feet.

The rack system will be six pallets high under a 24 foot clear roof height equipped with sprinklers and fire curtains. There will be a 12 inch back flue space in the rack system. There will also be in-rack sprinklers. The estimated equipment cost will be $100 per pallet position for the pallet racks or storage furniture and sprinklers. The cost of the in-rack sprinklers is included in the unit rack cost.

With a 4 inch spacing between pallets, and between the pallets and uprights, a pallet rack opening which is 56 inches wide will be required. Allowing a 6 inch clearance above the load, there will be 36 inches clear height between the four inch beams. This will make the load occupancy 40 inches high. There will be two pallets per beam with 4 inch spacing between them and between the pallets and the 4 inch uprights, spaced at 116 inches on centers. This results in a storage utilization unit of 70 cubic feet per pallet position with the 12 inch flue space. The actual full occupancy of this position will normally be 32 cubic feet of product, not including the plastic trays. We will use this as the "standard" unit load size. This gives us a 45.7% occupancy factor when the pallets are full. We will assume full pallet operations.

For costing purposes, and to include the pallet, we will use an occupancy of 40 cubic feet and a factor of 57% per pallet position as a standard.

Since we have applied conventional forklift equipment, we will assume the use of 2000 pound capacity, reach-type forklifts. These will require an aisle width of 8.5 feet with a 48 inch pallet load. We will also assume a lifting capability of 240 inches (20 feet).

The building "use rate" will be based on the same data as in the machine shop. The basic space cost "use rate" will be $0.012172 per square foot, per hour, including warehouse building burden expense.

However, in the case of storage operations, we use a cube "use rate" and cube occupancy charges. Therefore, if we assume a rack storage height capacity of 20 feet, the "use rate" per cubic foot will be:

$0.012172 per sq ft/hr/20 = $0.00061 per cubic foot per hour.

This computation is based on an eight-hour day which results in a "use rate" of $0.00488 per cubic foot per eight-hour day. However, since storage cost is normally computed on a 24-hour day, we will use this rate for the whole 24-hour day.

Using the above computed 70 cubic feet per pallet position, the daily "use rate" of a pallet position would be:

70 cu ft × $0.00488 per day = $0.3416 per pallet position per day

Sample Product Cost Calculations

Applying the 57% occupancy factor, the cube occupancy "use rate" per pallet of material being stored would be:

$0.3416/0.57 = $0.5993 per pallet per day

The "use rate" for the racks and storage furniture, their G & A charges, and the burden labor of the storage operation must be added to this.

Let us use the pallet stack "footprint," including rack clearances and half of the flue space, for the storage space computation. This defines a stack's building space occupancy to be 56 inches by 54 inches, or 21 square feet, plus 19.8 square feet for half of the facing 8.5 foot aisle. The total estimated space usage will then be 40.8 square feet per stack position.

This must be divided by 6 pallets high storage to compute an equivalent of 6.8 square feet of space per pallet position.

At $50 per square foot, this is $340 in capital investment. Adding the $100 rack, the pallet position investment is $440. Applying the previously computed G & A factor of $0.0000623 per dollar per hour, the daily general and administrative expense absorption per pallet position will be ($440 × $0.0000623 × 8) = $0.2193 per pallet per day. This must be added to the $0.5993 per day space "use rate" for a total of $0.8186 per pallet per day for storage.

In addition, we must capture the ownership cost of the storage furniture and equipment. Assuming a straight line depreciation rate of 10% (10 years), the annual cost of the racks will be $10 per pallet position or $10/1952 hr × 8 hr = $0.041 per pallet position, per day. This must be added to the previous value for an updated storage "use rate" of $0.8596 per pallet per day.

The cost of warehouse burden labor must also be added to this cube occupancy "use rate." In this sample demonstration case, we will assume a $0.050 per pallet per day burden labor cost for supervision, clerical, and maintenance staff. This charge will be computed using the same procedure as used in calculating the machine shop's burden labor. The resulting storage "use rate" will then be $0.9096 per pallet position per storage day.

In a non-palletized situation, the same procedure will be applied. However, instead of using a pallet unit as a cost module, a cubic foot of shelf space or a tray in a ministacker or other mechanized storage facility will be used as the "common denominator." It will be the basis for developing a standard time use of facilities "use rate" for storage cube cost.

The application of storage cost to the accumulated value of the part or product will be a function of the time that the item spends in storage. A daily "use rate" charge will be added to the inventory valuation of each item for each day the item remains in storage.

The interest cost of the dollars locked up in the inventory and the cost of inventory insurance should also be added to define the daily cost of ownership. This will be the only ownership cost element which is related to the value of the goods. These costs can be computed and added at the time the item is withdrawn from inventory.

This procedure will encourage inventory reduction and rapid throughput and turnover. It can also be used as a possible measure of the effectiveness of a Just-in-Time materials management program.

In addition to the storage "use rate" cost, there will be expenses for inbound handling, relocation of stock, order picking, and outbound handling. Computation of these costs will be based on standard handling times similar to those used in the calculation of the bar stock handling operations for the machine shop.

The typical pallet pick-up and placement times are 0.50 minutes for each operation and forklift trucks usually can average 200 feet per minute in travel. The time required for each "common denominator" move will be multiplied by the equipment's "use rate" plus the labor rate per minute to produce the operation or transaction cost. These costs will be added to the storage "use rate" for each item placed in storage or removed for shipment or use.

In addition, a typical order picking time is 0.08 minutes per carton for case picking and 0.10 minutes per piece for large loose parts. These times, when multiplied by the labor rate per minute, will be applied to each unit picked. The resulting cost will be included in the "landed cost" of the item at its next point of use or at the shipping dock.

Thus, the cost of warehousing the work-in-process, raw materials, and supplies used in manufacturing, and the resulting finished goods, will be the time use cost of the cubic feet of "housing" occupied by the materials. This will be computed on the basis of the "use rate" cost of the cubic space that the material occupied and the time it spent in storage.

In addition, the warehousing cost will include the cost of in and out movement and in-storage handling of the goods. These costs will be added to the manufacturing costs to produce the "landed cost" of the parts and products at their destination in the factory or at the shipping dock.

L. ACCUMULATING LEVEL-BY-LEVEL INVENTORY VALUATIONS

The key to accumulation of product costs on a level-by-level basis is the use of a level-by-level inventory or materials management system. In such a system, the items and their costs are recorded (posted) into inventory at each step in the process. To demonstrate this procedure, we will follow the production of the worm wheel and shaft sub-assembly from the steel bar stock and bronze ingots to the completion of the sub-assembly operation. In doing so, we will use the costs developed in the foregoing sections.

Raw materials and supply stores' inventory values for the components and materials required for the worm wheel sub-assembly are as follows:

Sample Product Cost Calculations

Raw Material Inventory Values:
   Bronze (06-025-0025), 64 lb @ $2.00/lb = $128/ingot
   2 inch diameter 1045 CRS steel rod, 22 ft (06-026-0027)
   22($0.70/lb × 10.7 lb/ft) or $7.49/ft = $164.78/bar

Purchased Component Inventory Values:
   Roller bearing (04-013-0014)   = $13.00 ea
   2 required   = $26.00/assy
   1/4 inch square keys (04-016-0018)   = $ 0.50 ea
   2 required   = $ 1.00/assy

The raw materials are withdrawn from inventory at their standard cost and used to make the semi-finished parts.

In the case of the bronze, the ingots are drawn from the inventory at $128 each and melted and poured into worm wheel castings. The castings leave the foundry with a new part number (03-009-0009). The bronze worm wheel castings weigh 44 pounds each and have a casting value of $95.00. They are recorded into the semi-finished parts inventory at that value. We are not adding a specific move cost here because the cost of general materials handling has been included as a burden cost in the manufacturing space's "use rate" and in the delivered value of the casting in the machine shop.

The two inch diameter, 1045 CRS steel rod is withdrawn from inventory in 22 foot bars (in bundles) at a value of $164.78 per bar. Since the shafts are made on a bar feed lathe, no slugs or semi-finished parts ever actually exist in an inventory. However, from a cost computation point of view, the material standard calls for fourteen inches per shaft. This standard material is recorded as a semi-finished part in the inventory system at a value of $8.74 per piece. Since it never really exists, it is immediately withdrawn and entered as material input into the cost of the finished shaft.

Semi-Finished Parts Inventory Values:
   Bronze gear casting (05-018-0023)
   44 lb @ $2.00/lb plus casting costs   = $95.00 ea
   Shaft slug–14 inches @ $7.49/ft   = $ 8.74 ea

These semi-finished parts are withdrawn from inventory and their values are entered into the costing of the machining operations as the raw material for those operations. In the case of the worm wheel, there are three levels of semi-finished parts to be posted into the inventory. The cost accumulation for the worm wheel at each level is:

Worm Wheel Blank Machining Operation:
   Material—bronze casting (05-018-0023) $ 95.00
   Machining—Turned worm wheel blank $ 7.524
                                                          $102.524/part

This worm wheel blank is entered into the work-in-process inventory and moved on to the hobbing operation as the input material for that process.

Worm Wheel Hobbing Operation:
    Material—worm wheel blank (05-018-0123) $102.524
    Machining—hobbed worm wheel gear $ 6.80
                                                   $109.324/part

This operation again produces a semi-finished part which must be posted into the inventory at the accrued $109.324 value. It may be immediately moved to the next work station or be placed into the WIP inventory. The hobbed worm wheel gear (05-018-0223) must next have its key way broached. Therefore, the hobbed worm wheel gear is the material for the broaching operation at $109.324 per piece.

In both of these moves, we will assume that the gear rested in WIP inventory for one day and accrued an additional handling and storage cost of $0.8636 per pallet per day for the two visits to WIP plus $1.00 for each in and out move. The total WIP storage and handling cost will be $3.727. This will make its entry value into the broaching operation $109.324 + $3.727 = $113.051. We earlier assumed that the cost of the broaching operation would be a standard $4.00 per part. This will make the finished part value of the worm wheel gear (03-009-0009) $117.051 as it enters the finished parts store room prior to assembly.

In the case of the worm wheel shaft, all of the operations are performed on the same machine and no semi-finished parts are created. However, we enter the standard CRS steel slug (06-026-0027) value into the machining operation as a withdrawal from WIP inventory and as the unit of material for the turning operation at a value of $8.74.

Worm Wheel Shaft Turning Operation:
    Material—2 in CRS 1045 slug $ 8.74
    Machining and handling $ 7.179
    Finished worm wheel shaft $15.919/part

The finished worm wheel shaft (03-010-0010) is entered into the assembly store room inventory as a finished part at the $15.919 value.

Two worm wheel spacers (03-012-0012) are also required. The cost of these parts was estimated to be $3.00 each. They are also in the finished parts inventory.

The assembly store room inventory also contains some purchased components for this sub-assembly. They are:

    Roller bearing (04-013-0014) = $13.00 ea, 2 required   = $26.00/assy
    1/4 inch square keys (04-016-0018) = $ 0.50 ea,
        2 required   = $ 1.00/assy

Sample Product Cost Calculations

All of these parts are next withdrawn from the finished parts inventory as material for the assembly operation. The total of the materials for assembling the worm wheel and shaft sub-assembly is:

Worm wheel shaft (03-010-0010)	= $ 15.919
Bronze worm wheel (03-009-0009)	= $117.051
Two warm wheel spacers (03-012-0012)	= $ 6.00
Two roller bearings (04-013-0014)	= $ 26.00
One bearing seal (04-014-0015)	= $ 2.00
Two 1/4 inch square keys (04-016-0018)	= $ 1.00
	$167.97

The cost of the assembly operation is added to the material cost to develop the finished sub-assembly cost as follows:

Worm wheel assy (02-004-0004) materials	= $167.97/assy
Assembly operation—2.80 minutes	= $ 2.968/assy
Finished sub-assembly value	= $170.938/assy

The completed worm gear and shaft sub-assembly will enter the WIP inventory as part number (02-004-0004). It will then become part of the "material" used in the speed reducer final assembly operation at an inventory value of $170.938 plus accumulated storage costs.

Thus we can see that the system accumulates material, time use of facilities "use rate" charges, handling expenses, and storage costs as the part progresses from one inventory level to the next through the manufacturing process. The incremental operation cost is added into the item's inventory valuation at each level and the fully absorbed value of the part or sub-assembly can be defined at any point in the manufacturing process.

This procedure also defines the sources of the costs in the manner of "activity based costing." It recognizes and defines the costs generated at each work station and operation, and for each element of each operation and each period of storage.

This procedure will also allow the definition of the cost of partially completed parts for sale to sub-contractors as material for their supporting operations, or the use of semi-finished parts as the material input to the manufacture of a modified version of the part.

For example, a machined worm wheel blank (05-018-0123) could be sent to hobbing as the material to make a specialized gear with a different tooth pattern. Alternatively, a hobbed worm wheel gear may be used as part number 05-018-0223 without a broached key way and be shrink mounted on the worm wheel shaft. This would create a different sub-assembly and sub-assembly part number.

This level-by-level costing procedure also supports the use of parts in the after market service parts operation. It provides a definition of the manufacturing cost of each part and permits the service parts marketing function to set a realistic replacement parts price.

In any case, the precise accrued cost of each part will be available for computation of the assembled cost of the speed reducer and any of its variations.

M. DEFINING THE "TOTALLY ABSORBED PRODUCT COST"

The objective of the "precision costing" approach is to capture all direct and indirect manufacturing costs and to fully absorb them into a definition of the product cost without succumbing to the cost allocation errors imposed by variations in production volume or ratio costing on the basis of such variable bases as labor and material.

To be useful in manufacturing management, operational control, cost estimating, cost engineering, and product pricing, the product cost must be defined in three ways.

One needs to know the standard cost, the actual cost, and the landed cost at either the manufacturer's shipping dock or the customer's receiving dock.

As stated in earlier chapters, the actual cost of the product must be defined independently of the volume of production in the plant. The cost of idle capacity or the benefits of overutilization of the facilities are a function of marketing and sales effectiveness. They are not the result or responsiblity of the manufacturing operation. The cost of manufacturing the product is a function of the time use of the facility and labor plus the cost of the materials used.

The actual time use of the facilities or work stations involved in the manufacture of each product and its parts and assemblies provides the basis for applying the appropriate "use rate" to each work station. The "use rate" defines the time cost of using the factory facilities and the indirect burden and overhead expense which is chargeable to the product. It defines the period and variable indirect costs of an "activity" or operation.

The actual time also records the labor time spent in producing and handling the product. By applying the complete labor cost to the operation time, a second element of the cost is defined. Adding the cost of the materials used in the process makes the product cost accrual complete. These costs provide a realistic, non-production volume related, "activity based cost" for each operation and product.

These computations can use either standard or actual times. The standard times will result in a definition of the expected or planned totally absorbed standard cost of the product. It is the basis for cost estimating, cost engineering, process evaluation, and value engineering. If the standard is correctly predicted, it may not result in any product cost variances based on the actual cost of manufacturing.

Sample Product Cost Calculations

The actual time can be recorded by various timekeeping methods. Standard times are usually included in the operations' sheets or instruction documents or in the controlling computer programs. In any case, the cost of the product should be computed on the basis of both the actual and the standard times to capture and measure the variances.

The cost of production can be defined by recording the three key elements of the manufacturing process which are material, labor, and the facility "use rate," based on the time it takes to make the part or product.

For our example product costing we will use the standard times and costs computed above and some estimated parts' costs which were not computed to save space.

Finished Product
Worm Gear Speed Reducer	01-001-0001	$336.68
Housing Assembly	02-002-0002	$ 55.89
Housing	03-002-0005	$ 21.49
Front cover plate	03-006-0006	$ 7.50
Solid end plate	03-006-0007	$ 8.25
Seal hole end plate	03-006-0008	$ 8.25
16 Hex head bolts	04-015-0016	$ 9.60
Two allen socket plugs	04-015-0017	$ 0.80
Worm Wheel Assembly	02-004-0004	$164.01
Worm wheel shaft	03-010-0010	$ 15.92
Bronze worm wheel gear	03-009-0009	$117.05
Two worm wheel spacers	03-012-0012	$ 6.00
Two roller bearings	04-013-0014	$ 26.00
One bearing seal	04-014-0015	$ 2.00
Two ¼ inch square keys	04-016-0018	$ 1.00
Worm and Shaft Assembly	02-003-0003	$ 83.15
One Piece Worm Shaft	03-003-0011	$ 33.65
Two Ball Bearings	04-013-0013	$ 47.00
One Bearing Seal	04-014-0015	$ 2.00
One ¼ inch square key2	04-016-0018	$ 0.50
WIP Storage & Handling -Estimated-		$ 14.00
Assembly Operations-Complete Reducer		$ 15.90
COMPLETED REDUCER		$336.68

Thus, the totally absorbed cost of the worm wheel speed reducer is computed to be $336.68 based on the time use of facility "use rate," the standard labor content, and the standard materials. This cost estimate is not dependent upon the volume of the plant's manufacturing operations.

The cost of unused capacity and/or the benefits of absorbing burden through overutilization of facilities are recognized in the profit and loss statement of the enterprise. They do not affect the computed unit cost of the product. The "for nothing" costs resulting from overtime, holiday, or multishift operations are also treated as a general overhead expense resulting from scheduling. They are fed to the product cost on a period basis via periodic revisions in the "use rate" of the work station and the whole facility.

17

Operations Analysis and Cost Variances

A. COST VARIANCES—A MANAGEMENT ANALYSIS AND COST CONTROL TOOL

The basic objective of a standard time or cost management system for management, scheduling, and control in manufacturing and warehousing is to provide a "baseline" against which to measure any deviations from predicted or expected performance. If the standard accurately portrays the planned work pattern, the deviations or "variances" from the standard will be random and minor. If the variances are skewed in one direction, either the standards are in error or the performance of the operation has changed or is in error.

The size of the variance defines the degree of the deviation from the plan. A skewed variance is the signal for corrective action in the operation or adjustment of the standard.

The standard costing philosophy presented in this book is based on the concept of a *total absorption standard cost system* that is independent of the volume of production. The elements of the system are the following:

- A *"time use of facilities"* standard cost which fully absorbs all space, equipment, and support area ownership costs, and all period, burden, indirect, overhead, and general and administrative expenses for use as a work station time "use rate" or "rental" charge.
- A *standard labor time cost* for each task, operation, and operator which includes all fringe costs.

- A *standard design material cost* for each part or component which includes predicted chip and scrap losses, material acquisition and ownership costs, and the effect of "nest and gain" or material scheduling decisions.
- *A level-by-level inventory system and numerically coded, matrix type, level-by-level manufacturing bill of materials* to accumulate all costs as they accrue during the process.
- A *matrix type part numbering system* which permits level-by-level accumulation of accrued costs into each item in the work-in-process and finished goods inventory as it flows through each operation in the system.

This costing philosophy is based on combining the standard cycle time "use rate" or rental cost earned by a work station in the performance of a manufacturing or warehousing operation with the standard time cost of each operation's labor, and the standard design cost of materials. These combined standards are the basis for developing the true manufacturing and warehousing operations' costs. These standard costs are independent of the facility's production volume or capacity.

The basic premise of the "time use of facility" approach is that the product should not be charged with the cost of unused plant capacity or benefit from overabsorbed period and overhead expenses.

Therefore, the standard cost of each operation, on each work center, is based on the measured actual or synthesized standard cycle time extended to include the "use rate" and labor expenses consumed on that work station during that operation only. The standard cost of making the part or product is the sum of these standard operations costs for all required operations plus the standard cost of the materials used and any scheduled WIP storage and handling costs. The accumulated standard cost for each operation is posted into the level-by-level inventory.

The actual cost of each operation is computed parallel to the standard cost. The actual cost is based on the sum of the actual consumed times for each operation in the process multiplied by the standard "use rate" and labor cost of each operation. The actual cost of work-in-process handling and storage, plus the actual cost of the material consumed is added to the computed actual cost of each operation to develop the actual cost of the part or product. This actual cost is also posted to the WIP and finished goods inventories in a separate column as it is accrued. At each posting or operation, the difference between the standard and actual cost is computed to define the variance for that operation.

The variances can also be computed for each element or component of the cost to focus on the specific cause of the variance. Total operation or process variances, or variances in facilities "use rate," labor, material, or storage

Operations Analysis and Cost Variances

and handling can be extracted and analyzed for each part, item, production lot, customer, department, work station, or worker.

A basic assumption in the application of the time use of facilities concept of precision manufacturing costing states that it is not essential to precisely balance operations costing and financial accounting. However, the "use rate" for each operation and part or product accumulates all of the indirect and period costs of the operation for each unit of production. Therefore, the computed standard and actual product costs can be used to calculate and project the break even point for a product, or for a whole manufacturing operation.

These same production floor source data are used by the financial and operations accounting functions. These data are derived from the measurement and recording of both productive and supporting shop floor activities. However, their use and analysis, and the resulting reports, can be designed to serve separate purposes and address different audiences. They do not necessarily need to be reconcilable to provide management with the required information and criteria for sound decision making.

Modern data base type computer software is available to support automated data collection, production and inventory management, cost accounting, and operations analysis. This software allows management to accumulate detailed manufacturing and materials management information and to manipulate it into a variety of report formats for different management audiences and purposes.

Manufacturing and engineering executives need precise and detailed standard design data and level-by-level operating cost information. These data are used to evaluate design and operational options, to help make product design and manufacturing methods decisions, and to estimate product costs.

Marketing executives need valid product cost estimates and true manufacturing cost data to determine the enterprise's ability to produce and deliver products within competitive pricing patterns and with acceptable profit margins.

The financial officer requires good data to prepare operating budgets, define and project cash flows, compute and evaluate return on investment, forecast and measure profit performance, and project tax liabilities.

All of these analysis requirements need and use the same production source data! Computers can manipulate these data and produce the appropriate report formats to meet each of these needs. Although the resulting numbers may not necessarily balance, they will provide the required decision making data for each management function.

Concurrently, the precision time use of facility, total absorption standard cost system will provide *precise, volume independent, manufacturing and product cost data* for making operating and engineering decisions and performing realistic pricing analyses.

B. OVER- AND UNDER-UTILIZATION OF FACILITIES

The philosophical basis for the time use of facilities' costing concept requires separate and definitive cost treatment of over or under-utilization of capacity. This approach requires the assignment of unabsorbed indirect expenses, and any profits from overabsorption of burden and overhead, to the profit and loss statement of the enterprise, and from there, to the business function, activity, or organizational entity which caused the deviation from the standard or full facility utilization.

At the same time, these variances must be defined in relation to the part, product, and operation which generated them.

Conventional cost accounting redistributes the over or underabsorbed, fixed and variable indirect, burden, and overhead costs of the business to the product in proportion to the level or volume of production. The time use of facility method assigns these schedule-based cost variances to the burden expense accounts of the enterprise's profit and loss statement. They can later be redistributed to the product via the space "use rate" on a period basis and in proportion to the product's time use of the facility.

This technique produces a stable, responsibility or activity oriented, standard definition of the facility utilization cost for each operation and part in the manufacturing system. It also provides a means for measuring variances in facility performance. The impact of volume and schedule will generate a variance in the relationship between the accumulated "use rate" charges and the total indirect cost of operations. This procedure avoids volume based distortions in the computation of product cost.

By establishing a standard "use rate" for each work station or activity in the facility, the utilization cost of the facility can be defined on a time basis, whether productive, idle, or performing in excess of plan. In schedule conforming operations, the "use rate" charges are assigned to the product on a time basis through the work-in-process costing procedure.

The actual time's "use rate" cost for facility utilization is posted to the inventory record. When the work station or facility is idle, its "use rate" is charged to an "idle facility cost" subsection of the factory burden account. This isolates these costs. The cost of idle facility is a definable expense in the profit and loss statement which does not falsely blame the factory for sales and marketing's failure to provide work. It also helps in responsibility accounting by relating idle capacity costs to the failure to develop new products or sell all of the plant's capacity.

Another factor which will cause a facility utilization variance is a change in the operation's cycle time. An increased "use rate" or labor cost variance might result from a poor operator performance or an equipment failure. Reduced utilization time could result from a more effective operator performance.

Operations Analysis and Cost Variances

In that case, the variance is chargeable to the product's actual cost and is entered into the inventory's actual valuation of the product. This does not alter the standard cost value of the product.

When production performance improves and underutilizes the facility because of unplanned time savings, it also defines the beneficial value of efficiency. In this case, the variance would show as the difference between a lower actual time and the predicted standards, and their cost effect.

In the case of overutilization, the "use rate" will be properly charged to each unit of the product to capture the true time use of the facilities. The variance in the total manufacturing facility "use rate" will include an overabsorption of the indirect burden and overhead expenses. These apparently overabsorbed costs will be entered as a cost reduction in the "idle facility cost" subsection of the factory burden account. They will reduce the unabsorbed burden cost and will help the profit side of the profit and loss statement.

This variance should also be averaged over the year and be a part of the calculation of the burden costs which are redistributed to the "use rate" on a period basis and thereby to the product. If the improvement in efficiency is permanent, and the variances are stable, the standards should be modified to include their effect.

Although this approach isolates the product cost definition from the variances in production volume, it defines the cost effect of changes in the schedule and facility utilization. The average annual cost of these variances should be periodically fed back into the product via the space "use rate" for the plant.

The value of each element of the work station's "use rate" variance should be defined so that it can be related to the product, lot, work station, and operator for management evaluation and/or corrective action. The value of the variance is also transferred into the burden account and redistributed to the work station's space "use rate" on an annual average basis (Figure 7.12).

C. DEALING WITH LABOR COST VARIANCES

There are at least three types of labor cost variances which are common in a manufacturing operation:

1. A variance in the operation's time cycle which makes the actual time and labor cost differ from the standard time and cost.
2. Assignment of an operator with a different pay grade (usually higher) to the task. This can result in a variance between the actual wages paid and the standard labor cost predicted for the operation.
3. Performing the operation on an overtime, night shift, or holiday/Sunday schedule which results in the addition of a "for nothing" cost.

In situation number one, the work cycle may be longer because of operator inefficiency, equipment failure, or material problems. The actual cost of labor will be posted into the inventory value of the product or work-in-process. The variance will be treated as an increased manufacturing cost and inventory cost variance.

The extended cycle time will also increase the work station's "use rate" charge to the operation. The actual cost will be posted to the inventory valuation and the variance from standard will be defined for both labor and "use rate" costs. This should lead to corrective action by supervision, but usually no change in the task standard.

If the variance results from improved operator skill or efficiency, or from a specific operator's unique skill, and the cycle time is shorter, two courses of action are possible.

If an incentive program is in place and the shorter time is specific to an individual operator or work team, incentive payments will be made and the cost variance will be noted in the actual cost column of the inventory record. No change will be made in the operation or standard in that case.

However, if the cycle time reduction (or increase) is consistent with all operators, the task standard should be changed. In the case of a reduction, suitable recognition should be given to the operators. In this situation, both the labor and work station "use rate" standards will be changed to the new time standard.

With a stable work force, a change in the operation's cycle time will not affect the total labor payroll cost unless an incentive bonus is paid to the efficient worker. This incentive cost will not be added to the standard cost of the product. It will show up in the actual inventory cost as a labor cost variance for the specific product lot.

The incentive pay, and any excess labor time variance, will be posted into the factory burden account for annualization and redistribution. All of the incentive bonus costs will be added together in the burden account. The average annual bonus cost for the plant will be computed. These costs will be distributed to the cost of manufacturing via the space "use rate" in the same manner as the cost variances in the use of the facility.

In labor variance number two, the change is based on labor policy and worker assignment. In most companies, and in most labor unions, assignment of a low paid worker to a higher paid job causes the worker to be paid at the higher wage rate for the assigned task. Conversely, if a high paid worker is assigned to a lower rated job, he normally will continue to receive his regular pay rate. As a result, we have two kinds of labor variances growing out of employee assignments. The operation cost will vary and the overall payroll cost may also vary.

If a low paid operator is assigned to a higher rated task, the labor cost for the operation will not change but the overall payroll will increase. There-

Operations Analysis and Cost Variances

fore, there will be a labor cost variance to put into the burden account but no change in the inventory value of the product or part. This variance will be distributed to the product via the space "use rate" in the same way as additional facility usage costs and incentive pay.

If the high paid operator is assigned to a low pay job, there will be an increase in the labor cost for the operation. This will result in a variance in the item's actual inventory value. However, there will be no change in the overall payroll cost. The operation cost variance will enter into the burden variance account.

The third cause of labor cost variance is based on the "for nothing" cost and its impact. This is the cost of scheduling. The author believes that the "for nothing" cost is a burden expense. It should be periodically allocated to the "use rate."It should not be charged to the individual product or order unless it is caused by a specific customer's scheduling requirement. In that case, the customer can and should be billed for the differential in labor cost. The standard cost or inventory value of the product should not be altered by the "for nothing" cost.

As stated earlier, if overtime is regularly scheduled, and if night shift operations are normal, the overtime and night shift bonuses or "for nothing" costs should be prorated over the total scheduled hours per week and the adjusted labor rate should be the standard labor cost. In that case, the "for nothing" cost is absorbed into the product as a normal part of the labor charge.

However, in the case of irregular overtime, holiday work, and night operations, the bonus or "for nothing" cost should be treated as a labor cost variance and be entered into the burden expense account. It is the cost of scheduling. It should not be charged to each operation, but should be redistributed into the product via the space "use rate" based on the average total period "for nothing" cost for the whole enterprise or plant.

In summary, variances in labor time and cost should be recognized in the actual item and operation costs posted to the inventory value of the part or product. These costs should also be entered into the burden expense account. Annual adjustments in the space "use rate" should recognize their impact. All "for nothing" costs should also be entered into the burden expense account and be fed into the product cost via the space "use rate."Only those "for nothing" costs which are customer induced should be assigned to a product and charged to the customer.

D. EQUIPMENT SELECTION AND UTILIZATION VARIANCES

It is obvious that the choice of equipment or machinery will affect the manufacturing method, the time required to perform an operation, the capital investment in the work station, and the space required. Each of these factors will have an impact on the "use rate" of the work station and the unit cost of each part made there. In a parallel effect, the choice of machine will define the required skill level and have an impact on the wage rate of the operator and the labor time.

The choice of equipment can be made at several points in the planning system. In the initial design of the product and process, a machine or set of equipment is usually selected to define the process and assure the manufacturability of the part or product. This usually becomes the "standard" for the process at a projected production rate. The standard cost of each operation is then computed on the basis of this work station's "use rate."

In some cases, more than one "standard" might be selected in order to deal with differences in production rates or volumes. For example, if only one unit of a product is to be made, it might be wise to accept the labor cost required to make it on standard or conventional machinery and eliminate the cost and time required to program and tool a CNC unit. This decision would reduce the "use rate" cost by reducing the capital invested in the work station, but increase the labor and "use rate" time content of the operation, and probably lower the employee wage rate.

By comparison, if a large number of pieces are planned, the schedule may suggest a CNC unit. This would involve a higher capital investment, more setup cost, and a higher skill level wage rate. This would result in a higher work station "use rate" charge. However, since the time cycle would be reduced, the unit cost may be much less. The combination of these decisions may result in an increase or decrease in the operation cost. This would affect the inventory value of the part being produced.

The variance could be based on the cost of making one piece on conventional equipment versus manufacturing the part on a CNC machine. The choice of the "standard" machine would determine the character of the standard cost and the variances which might result from the scheduling of the process.

If the CNC machine is the standard, the cost of making one piece on that machine must be compared with the cost of using the conventional work station. In any case, the part cost variance would be captured by comparing the actual cost inventory valuation with the predicted standard cost.

In addition, there would be variances in the labor cost based on the different wage rates and process times. These variances would show up in the details of the manufacturing cost, but not usually in the payroll account.

Another equipment based cause of part cost variance would be the use of a different machine or process from the standard to make the part. For example, the above described 8 inch pitch diameter worm wheel casting could be machined on a 36 inch Bullard vertical lathe as opposed to either a conventional engine lathe or the proposed CNC turn/mill center. The decision to do so might be based on machine availability and a need to make the part without delay.

Operations Analysis and Cost Variances

Such a scheduling decision would result in a different work station "use rate," a different labor rate, and a different work cycle time. Each of these elements, and the whole operation, would generate a variance in the actual versus standard cost of the part. In most cases, the total labor payroll would not have any variances and the material cost would not usually be affected. In some cases, the variances might even compensate one another and the total cost of the part may be the same.

In summary, the choice of the "standard" machine or work station defines the base against which the actual costs are measured. The choice of the machinery and equipment can affect the time cycle of the operation, the skill and wage rate of the operator, and the "use rate" charge for the work station. The choice of the work station or machine determines the basic or standard cost of the operation.

The variance in the product or part cost is the difference between the standard work station and operation cost, and the alternative work station or performance cost. This variance is part of the actual inventory valuation of the part or product. It will also be posted to the variance subsection of the burden account and be periodically applied in an adjustment of the "use rate."

E. DEALING WITH PROCESS VARIANCES

There are several types of process variances. Most of them involve changes in the design of the product or in the choice of materials. The need to modify a product's appearance to meet "private label" requirements might also cause a change in the final assembly or finishing process. In some cases, the choice of the method of manufacture might be the result of the order quantity required.

The most likely process variance is the result of a planned alternative process based on the economics of scale or quantity in the lot size to be manufactured.For example, if only one of the worm speed reducers discussed above is to be made, it might be more economical for the housing to be a weldment rather than a set of castings which would require the production of patterns and molds. However, if a significant number of units are required, the casting process may be the most economical.

If a company makes the product in response to customer orders rather than producing them to inventory, these manufacturing process alternatives might be a decision issue at the time of releasing the customer order into the factory for manufacture. Any variance in the cost of the housing would affect the sales margin, and in some cases, the price quoted to the customer. The process decision might need to be made at the time of the sale. If a different process is later used on the order, a process variance would develop.

Again, referring to the speed reducer, the worm thread might be made by rolling instead of turning in very high-volume operations, and be turned when only a few are required. This option might significantly alter the machining cost of the worm shaft and ultimately, the cost of the finished product.

In each of these cases, the alternative process would have a predicted "standard" cost. The alternative part design would have a distinct part number based on an engineering change or configuration control suffix. It would have a different standard cost in the alternative bill of materials. The standard cost would be based upon the "use rates" of the work stations and the wage rates of the workers expected to be used. If one of the processes is selected as the standard, the process cost variance would appear in the sub-assembly and finished product inventory valuations.

Another process variance might result from the choice of different materials for a particular production run. In some consumer products, gears and housings can be made from either die castings or plastics with equal performance capability. In the case of gears, the use of plastics such as nylon can eliminate machining and preclude the need for lubrication. Plastic gears are also quieter, but they may be less durable and cause service problems in the field. In the case of housings, one advantage of the plastic version is the elimination of the painting operation through the use of colored material. The disadvantage is needing to inventory a multiple of WIP housings rather than adding color to a standard housing at the end of the assembly process.

Again, each alternative part will have a distinct part number. The resulting product or sub-assembly will also have a different part number and bill of materials. This will result in different process standard costs, different part costs, different sub-assembly and final assembly costs, and different WIP inventory values. The variances will be in the sub-assemblies and finished goods. The parts will be distinct items and have their own standard costs.

In the case of private label differences, the variances will usually appear in the final assembly and packaging processes as a result of labelling and packaging to customer specifications. These process variances will impact the final product and its assembly, but should not affect the detail parts' costs.

In cases where non-standard operations and processes occur in the manufacture of otherwise standard assemblies, the variances will be treated in the same way as in parts manufacturing. The actual cost will be compared to the standard in the work-in-process inventory to define the variance. In all cases, whether the variance is in a part's cost, a process cost, or a process variance in the sub-assembly or final product, the variance will be posted to the factory burden account for redistribution to the product through the space "use rate" on a periodic basis. It would also be handled as a profit and loss item.

The alternative process decisions can be rules driven or artificial intelligence driven if precision standard costs are in the system. The resulting variances will be defined in the WIP inventory valuation and in the costing of the finished product.

F. DEALING WITH MATERIAL COST VARIANCES

Material cost variances were discussed as a part of the material standard cost development. The scrap, "nest and gain," and shrinkage adjustments create variances between the standard material cost and the actual material used. But these are predictable quantitative adjustments which can be included in the standard cost computation or redistributed to the product via the space "use rate."

The true variances in material costs are usually based on changes in the price of the material over time. If the inventory system uses "first-in, first-out" (FIFO), costing, an inflationary material price trend will result in a gradual increase in the actual material cost compared to the standard cost as time progresses. The "last-in, first-out" (LIFO) inventory costing will also result in an increasing variance, but probably at a more rapid rate. If there is a declining price trend, the variance will tend to shrink in both the FIFO and LIFO systems and could become a negative variance. These price based material cost variances will show up in the material inventory valuation as shipments are received and posted. These variances should be shifted to the main material account as miscellaneous material cost adjustments. They should be redistributed to all of the product's material as a periodic adjustment and applied in a manner similar to the shrinkage and "nest and gain" allowance.

Thus, the materials cost variances should be absorbed within the materials accounts and should respond to the market trends with periodic (annual or semi-annual) product based material cost standards' adjustments.

G. PRODUCT COST VARIANCES AND THEIR CAUSES

Product cost variances are caused by a complex set of cost element variances which can be compensating and invisible in the finished product cost.

For example, the cost of materials can vary with the market and labor cost can increase as a result of union action, legislation, taxes, fringe benefit costs, and employee performance. At the same time the cost of production can be decreased through improved operations (less time use of facility cost) and because of lower overhead and burden. The result can be no change in product cost.

Alternatively, the cost of materials could decline while the labor cost increases and the two could balance and cause no change in the finished product cost. In other scenarios, one rising cost element, such as labor or material, can cause a sharp increase in total product cost.

It is essential to monitor all of these cost elements and to manage them both individually and as a system. Figure 7.12 shows the means of accessing these cost elements for analysis and comparison via the SIMSCODER system. As shown, a variety of cost elements can be identified, valued, and compared by using these procedures.

Labor cost is accessible through the employee's clock or identification number. The material cost can be captured from the part number, the bill of materials, the operation sheet, and the WIP and material inventory valuation. The actual operation time can be recorded automatically by the machine or clocked in by the employee. It can then be compared to the planned standard time on the operation sheet. The correct "use rate" can be identified from the machine or work station number and the operation number, and the time use of facility can be captured from either the machine time or labor time.

All of these cost elements can be compared to standards by using an ongoing comparative analysis procedure within the inventory valuation process. These data can then be used to compute the actual and standard product costs and variances.

H. WAREHOUSING COST VARIANCES AND THEIR CAUSES

In a manufacturing environment, warehousing falls into three basic functional categories. They are the inbound raw materials, purchased parts, and supply stores warehouses; the work-in-process storage operations; and the finished goods and service parts distribution centers.

The calculation of standard and actual costs in each of these storage areas should be based on the cubic feet of material stored or moved, and the nature and volume of transactions, or movements in, out, and within the operation. The product's value or identity has little impact on the cost of storage or handling. The size and shape of the material and its relationship to the "common denominator" handling and storage system has a significant impact on the cost of the operation.

A key factor in the generation of storage and handling cost variances from standards is the variation in the number of pieces per "common denominator" handling and storage unit. The standard cube cost of handling or storing a "common denominator" pallet or tote box is usually the same, whether it is full, empty, or only partially full. The cube occupancy or "use rate" cost (in shelving and racks) is the same for the pallet or tote box regardless of its contents.

As stated earlier, the standard storage cost per piece, per storage day, is computed on the basis of a standard number of pieces per "common denominator" unit of handling and storage. Therefore, the actual cost of storing a partially loaded "common denominator" pallet will result in a plus variance in the cost of storing the product's pieces. The same kind of plus variances will develop from the handling of partially filled "common denominator" tote boxes or shelf locations.

The handling and storage of loose items will also generate a plus variance if the standard cost is based on the use of a "common denominator" handling system. This will not be the case when the handling of loose items without a "common denominator" is the standard method. In this situation, variances will be produced by variations in handling efficiency or by using non-standard handling methods. Conversely, if normally loose items are handled and/or stored in a full "common denominator" tote box or pallet, the handling and storage cost will be lower than standard and produce a negative variance.

Other warehousing variances can be found in the order picking and the shipping and receiving areas. If the usual sales quantity is a single unit, orders which require full cases or pallet loads will generate a negative variance in the order picking cost because of more efficient operations. Conversely, case pick or pallet pick items which are ordered in single units will produce a plus variance in picking costs.

Likewise, if less than pallet shipments (in or out) are the norm, palletized receipts or shipments will result in a lower unit cost and a negative variance. The reverse will also be true if normally palletized items are received or shipped in single units or cases.

Another cause for storage cost variances would be the placement of goods in other than the planned standard storage area. For example, the placement of canned goods in a refrigerated warehouse would generate a higher cube "use rate" cost than if they were stored in a standard or conventional dry warehouse. The floor stacking of pallets which are normally rack stored will also change the cube "use rate" and produce a storage time cost variance.

Labor cost variances are also probable. In most warehouses, there are different pay grades for fork lift operators, order pickers, dockhands, and other warehouse workers. However, with the exception of a few tightly unionized operations, workers will move freely from one job assignment to another. This will cause the same kinds of labor cost variances as were discussed in connection with the machine operators. In most cases, the worker will be paid the higher of the job assignment pay or his regular wage rate. This will cause variances in either the operation cost or the payroll expense, depending on the nature of the change in assignment and the wage rate of the worker.

Since warehousing is usually a non-cyclic operation, cost variances will be very common. In most cases they will average out and stabilize over time. An average annual net cost of the plus and minus variances can be applied to the "use rate" as a part of the warehouse's space or cube cost burden. This cost will then be applied to the item's warehousing and handling cost as a part of the standard.

At the same time, comparison of the standard and actual warehouse costs which are introduced into the product's inventory valuation will provide a detail measure of the warehouse's performance against standards.

Another warehousing cost variance will be generated by varying the dwell time of the product in storage. If the standard WIP or supply stores warehouse cost is based on a specific throughput rate, a longer or reduced stay in storage will alter the storage cost element of the process cost of the product. This variance can be used as a measure of the process' adherence to a Just-in-Time operation and/or an indication of inventory fluctuations and excess, or inadequate stocks in the material flow system.

I. INFORMATION FEEDBACK FOR COST CONTROL AND PRODUCT DESIGN

The primary reason for developing a precision, time use of facilities, standard operations and product costing system is to provide management with decision making information which is not tainted by tax law issues or production volume considerations.

The decision making functions which need these cost data are product design engineering, product and manufacturing cost estimating, manufacturing engineering and planning, manufacturing operations management, production control, inventory valuation accounting, financial management, and marketing and pricing.

In each case, the users need both the fully absorbed standard or expected cost of the product, parts, and operations, and the current cost variances and actual costs. The cost system must provide a means for the feedback of these data to the using activities.

In actual manufacturing operations, the time and piece count reporting at each operation is usually the medium for collecting both standard and actual cost data. In modern systems this is accomplished by using a combination of automated machine time reporting, labor work time reporting, and automated identification (bar coding, radio, etc.) reporting of product movement and inventory. Bar coding techniques are also used for worker time reporting, document to computer interfacing, and shop operations control. Wanding of a document bar code can bring up required data on a computer screen or initiate a machine operation program. It can also trigger the printing of required shop drawings, operations sheets, and other shop floor documents.

In state-of-the-art systems, the shop floor transaction data feeds directly into a computer data base. This data base makes the same detailed shop floor data available for use in design engineering, estimating, manufacturing engineering, manufacturing costing, production and inventory management, and financial accounting.

It is obvious that the computed standard or expected cost elements can also be published in the form of costed bills of materials, costed operations sheets, valued inventory reports, and price lists for purchased components. These data can be made available in either hard copy files or in the computer data base.

Operations Analysis and Cost Variances

When in hard copy form, these cost data are readily available to document users throughout the operation. This makes data access easy, but it also precludes good management information security. When these data are in a computer based system, they can be made instantly available to authorized users. At the same time, the use of passwords and limited access software programs can protect critical data from being accessed by unauthorized personnel.

In summary, by using the precision, time use of facility, total absorption, standard costing system, the central data base of the manufacturing management computer system will accumulate the required detailed standard data on all parts, materials, sub-assemblies, finished products and manufacturing and warehousing operations on an operation and level-by-level process and inventory basis. In addition, the shop floor reporting system will capture the actual time use of facilities and labor, and the inventory system will capture the actual material usage. All of the shop generated costs will be accumulated into the WIP inventory system and the product will be progressively costed and valued in the inventory system.

Accrued standard and actual costs will be available for each piece and operation at every step in the process. Whether in hard copy file form or in a computer data base, these data represent all elements of the costs generated by manufacturing activities. All elements of data, and the accrued summary data, can be fed back to those who require the data for planning and operating decisions, product and process design, and general management decision making.

The results of this precision costing procedure can be applied to the management of manufacturing and/or warehousing in either an activity based costing system or time use of facility standard costing procedure. In either case, the costs developed by these techniques are free of distortion by variations in production volumes or tax rules. They are a dependable standard against which to measure performance and define true product costs. These data can be made available in formal reports, published standards, or by direct access to computer files.

18
A Look into the Future

A. THE CHANGING INDUSTRIAL SOCIETY AND ECOSYSTEM

The "Industrial Revolution" continues! It has never ended! It progresses unabated throughout the world and will continue far into the future years and centuries. Manufacturing operations continue to be increasingly capital intensive, more productive, and less dependent upon people. High technology has spread throughout the world. Worker productivity is growing as a result of mechanization and automation. The dream of the people-free factory has almost been achieved in some modern industries.

For example, the share of American workers engaged in full time manufacturing employment declined from 22% in 1980 to 16% in 1993 and this percentage continues to shrink (from *U.S. News & World Report*, August 9, 1993). Also, "Output keeps climbing while jobs decline. That's a trend that will probably go on well into the next century . . . technology steadily displacing more-expensive human labor in our economy" (from *The Kiplinger Washington Letter*, August 27, 1993). Drs. Lowell Gallaway and Richard Vedder (Ohio University's Kennedy Distinguished Professors of Economics) suggest that "increasingly higher wages, and their replacement by investment in capital intensive production equipment, will cause increased unemployment and continuation of a permanent structurally unemployed segment of the population."

This decline in production employment is partially due to the export of American manufacturing jobs to low labor cost emerging countries. However, mechanization and automation are probably the major causes of increased productivity and the reduction in manufacturing labor content. This global trend has already created major changes in industrial economics and the world's socioeconomic structures. It will continue to do so!

Industrial technology has also increased the need for better educated factory workers and better educational systems. Capital intensive manufacturing has limited the opportunities for entry level and unskilled labor, and reduced the ratio of labor cost to material and burden costs. Mechanization and high wages have also eliminated, or redistributed, many of the human skill and muscle tasks from manufacturing and warehousing. "Just-in-Time" (JIT) manufacturing and distribution policies have reduced inventories and thereby reduced the need for warehousing and its unskilled and semi-skilled employment opportunities. Containerization has almost eliminated stevedores and similar transportation support jobs.

The world's economic system has also changed! Industry and marketing have become *global*. Large corporations are becoming truly international in their operations. Trade restrictions are weakening within and between regional market groupings. High technology has penetrated all cultures and societies. Modern industry operates in worldwide "Real Time" with computers, faxes, satellites, and air transport support.

The industrialized nations are approaching a no growth, or slow growth, replacement economy, status. Their own markets for basic consumer hard goods and discretionary consumer "big ticket" items are stabilizing or declining. Their standards of living have become mechanized and are nearing a quality of life plateau. The "poor" in the industrialized world often live far better on their minimum wages or welfare than much of the general population in areas such as India, Egypt, Black Africa, and Latin America. The poor in these areas are often destitute and without purchasing power.

At the same time, the developing nations are demanding, and producing, culturally focused goods for local and export markets. India and China build tractors, China makes bicycles, and Brazil makes automobiles. But Brazil, Indonesia, and China also build airplanes!

Most professional observers, industrial leaders, and politicians have recognized and commented upon the internationalization of large corporations. They have also criticized the movement of many key consumer and industrial goods manufacturing industries, jobs, and operations from the United States, Canada, and Western Europe into the Orient and other low-wage developing countries. However, these same critics have also failed to openly recognize the strong job-reducing impact of automation and mechanization in the factory, or computers and high technology equipment (such as FAX machines and photocopiers) in the office and middle management.

These technology based labor elimination trends in manufacturing, distribution, administration, and management do not appear to be fully recognized or understood by some key political leaders. These leaders either have not recognized, or have avoided acknowledgement of, the global technical revolution which has caused these changes. They have erroneously and publically attributed these trends to national or regional government political and economic policy.

Government has had very little influence on the development of these industrial trends. Global competition and market forces are the motivators!

Manufacturing industry is responding by investing in technology, laying off workers, internationalizing, and downsizing. They have implemented technology and moved plants overseas, but they have not faced up to, or dealt with, the critical socioeconomic and political impacts of these changes. These leaders have just begun to adjust their attitudes and policies to accommodate the impact of global industry and high technology on traditional and/or current North American and European management methods and policies. They are just beginning to recognize their global citizenship responsibilities for environmental management and the socioeconomic health of their host nations.

To date, Japan has led the world in recognizing and meeting the global market's challenge. They import almost all of their raw materials and convert them into the world market's requirements. They have taken over the consumer electronics, camera, watch, tableware, machine tools, and other markets in the industrialized world. They are a major factor in the automobile industry. Their motor bikes are a good example of their ability to capture a culturally focused consumer hard goods market. The Japanese have flooded the world with motor bikes and motor cycles and have preempted the other producers in Italy and the United States.

To meet these challenges, *management attitudes and practices must change*! Western management seems to have forgotten the basic fact that The true wealth producing activities of our world are the conversion of natural resources (minerals and agricultural products) into human usable and socially acceptable goods through the application of human labor and ingenuity for the benefit and comfort of people and society. When operating cost effectively, manufacturing produces the real wealth of nations and provides profits for the entrepreneurs. But, cost effective operations in a high tech world require precision costing and control.

One of the major problems in the present global economy is the spread of high technology and capital intensive manufacturing into the very low labor rate areas of the world. For example, India has many thousands of highly educated engineers and skilled craftsmen. India is also purposefully moving into computer managed manufacturing while continuing a pattern of dollar a day labor rates with no welfare system to support the unemployed poor.

The same technology is spreading in America and Western Europe, but with 10 to 25 dollar per hour labor rates, and in many cases, a shortage of engineers and technical personnel. The West also burdens industry with minimum wage laws and labor taxes to support their unemployment and welfare programs.

Another key problem in the international ecosystem is the dominance of the financial manipulators over the manufacturing industries. These people are reluctant to invest in modern manufacturing technology or new product research. They even cut research funding to improve stock holder dividends and short-term corporate profits. Their decision criteria are short-term return oriented. Financial manipulation produces nothing for society! It only provides wealth for the manipulators.

This situation relates to another capital investment problem. A critical manufacturing management practice includes the usual American policy of requiring a high return on investment (ROI hurdle) for approval of a capital investment. As manufacturing becomes more capital intensive, the validity of a three year or shorter ROI becomes increasingly questionable. The world's most successful manufacturing economies use a much longer period (often 10 years) for their capital recovery. They view their investment in automation as a competitive marketing tool. They don't insist on justifying each element of a system in isolation.

In capital intensive, high tech industry, most of the operating expense is attributable to the time use of the facilities and equipment used in the process, and the materials and supplies used in production. In capital intensive operations, direct labor is seldom linearly related to production volume or product cost and sometimes has little impact on manufacturing cost.

Automation investment recovery is seldom justifiable on the basis of labor savings alone, even in the high labor cost environment of the United States. One must recognize the competitive impact of automation through improved quality, reliability, customer responsiveness, and the ability to reduce inventory investment by applying a Just-in-Time management philosophy. More importantly, automation often provides the ability to continue doing business in the face of competition with low labor cost producers.

In spite of these trends, many companies continue to make their manufacturing, engineering, and marketing decisions on the basis of the old "generally accepted accounting principles" which are income tax and labor cost oriented, percentage based, production volume sensitive, and use a three year or shorter ROI hurdle. These costing and decision policies were developed before computers made precision costing practical.

It is time for a change!

In the rapidly changing global economy, manufacturing and warehousing operations managers and product design and development engineers need precise, detailed, level-by-level, and volume independent, cost data to make sound decisions. Precision costing of parts, products, and operations can help make high technology decisions more realistic and their results more predictable.

A Look into the Future

B. COSTING COMPUTER INTEGRATED MANUFACTURING (CIM)

The proponents of Computer Integrated Manufacturing (CIM) assure us that we will soon be seeing computer managed factories with very few workers on the factory floor. They tell us that robots and automated guided vehicles will replace assemblers, machine operators, and forklift drivers. They say that automated design retrieval and artificial intelligence systems will help us to design products and download their designs into tool design and machine operation software. They speak of Just-in-Time operations and Total Quality Management (TQM) as the foundation stones of CIM. These technical predictions are coming true.

However, few of these technological pioneers address the need to change management's financial bias and the old style cost accounting methods in order to properly address the capital intensive nature of CIM. If we are successful in the development of CIM, the conventional costing methods will distort and disguise the cost data from the factory floor. It is essential to convert the cost measurement procedures into a format which will recognize and capture the effects of capital intensive manufacturing facilities as the cost producing elements of the process.

The basis for burden and overhead cost recovery must be shifted from a labor or material base onto an invested capital base. The variations in the apparent product cost which result from the conventional volume based distribution accounting approach will not be acceptable in a capital intensive, Just-in-Time environment.

The time use of facilities or "use rate" costing concept provides an accounting tool which recognizes the capital intensive nature of CIM operations and properly measures performance.

In a CIM operation, the key management issue will be capital utilization. Labor costs will be spread over the whole operation in the form of computer programmers and engineers, setup technicians, and administrative staff. These labor costs can be treated as a burden on the capital equipment and facilities.

There will be a minimum of such direct labor as machine operators and materials handlers. The cost of material can be easily defined through the inventory control system.

As a result, it is essential to manage a CIM operation by defining the time use cost of the work stations and their equipment. The time use of facility "use rate" approach is designed to achieve this objective. It accumulates all of the indirect costs of the operation into the "use rate" of the work station and applies these costs to the product on a time use or rental basis. Any direct labor, and all material, is added to this time based rental or "use rate" charge to compute each operation's cost for the parts in process. The product should only be charged for its proper share of the use of the facility, materials, and direct labor. This computation method will properly define the true manufacturing cost of the product independently of the variations in the volume of production.

The overhead and burden cost of idle facility capacity is not a manufacturing cost!

Idle capital facility capacity cannot be "laid off" but it continues to generate the costs of ownership. These costs should not be charged to the product. The lack of work is usually a result of the lack of sales to fill the capacity. It is a function of marketing and new product development performance, and is the responsibility of enterprise and marketing management, not manufacturing. *But, the cost must be accounted for in the profit and loss accounts of the enterprise.*

The overabsorption of facility expense is usually a result of manufacturing scheduling to support good sales results. This gain or profit is also attributable to marketing performance.

Both of these cost variances should be treated as profit and loss adjustments and be credited or charged to marketing or product development operations. They should not be treated as manufacturing expenses or cost savings.

Having addressed the cost definition issue by applying the "use rate" time use of facility production costing technique, it is now necessary to relate this precision cost system to the management of a CIM manufacturing operation.

One of the obvious features of CIM system management or computer-managed and mechanized operations is their precision scheduling requirements. The machine time cycles and the automated materials handling systems dominate the scheduling process. They are precise and predictable, and normally do not vary once established.

Also, most of the limited labor content of the system is internal to the machines' work cycles and its cost is buried within the work centers' "use rates."

Because of these factors, the manufacturing operations and the shop loading process will usually be dominated by the machine or work station cycle times and their workload scheduling requirements. The scheduler will attempt to fully load the facility capacity and to minimize idle work station time. The scheduler will also try to schedule the workload within the standard working hours of the plant in order to minimize extra supervisory overhead and "for nothing costs."

This approach to shop loading and manufacturing cost management is equally applicable to conventional and capital intensive manufacturing operations. The shop is loaded with the objective of optimizing its capacity utilization in each case.

Since the machine or work station cycle time is the dominant factor in scheduling, it is logical to use it as the dominant element in the cost computation. The definition of the time use of facility "use rate" for each work station

A Look into the Future

includes and fully absorbs all of the overhead, burden, general and administrative, and period costs of the manufacturing operation. This full absorption cost accumulation allows us to precisely and fully define the cost of each minute or second of each work station's operation.

By defining the work cycle time for scheduling, we also provide the data for defining the work station's "use rate" cost for the operation. When the direct labor and material costs are added, we will have the data basis for developing the precise cost of each operation and part in the manufacturing process.

Having defined these cost elements, it is necessary to collect and compare the actual time and costs against the pre-calculated standard costs and schedules. In modern operations this is often accomplished by having the machines automatically report each work cycle to a computer data base. Another technique uses bar code wanding by the operators or materials handlers, or the inclusion and querying of a radio (RF) transponder in the "common denominator" tote or pallet load unit which transports the work-in-process (WIP) between operations. In any case, the controlling variable is time, and the computation of variances is based on the multiplication of the actual process time by the standard costs in the data base.

As discussed above, these costs and variances are entered into the WIP inventory record to produce a cumulative running account of the progressive cost of manufacturing the product. This can be accomplished in either a computer or a manual system. In a CIM operation we would use a computer and the cost and schedule data would be meshed with the materials requirements planning (MRP) system and the enterprise's (MIS) management information system.

C. CAPITAL INTENSIVE INDUSTRY REQUIRES PRECISE COSTS

When management invests in a CNC type machine, the cost is often several hundreds of thousands of dollars. When considering a Computer Integrated factory or a flexible manufacturing system (FMS) the multimillion dollar investment requires thorough research and sound economic projections. The high-volume production of low-cost items using "hard automation" also requires a big investment and precise cost predictions.

These investment cost projections require good market research data, thoroughly detailed product and process design information, and precise manufacturing costs. Even the largest corporations cannot accept multimillion dollar capital investment errors or inaccurate product manufacturing cost estimates.

Capital investments are usually "one shot" expenditures and cannot be recovered without profitable production. Unlike labor intensive operations, it is difficult to "lay off" an automated manufacturing system. Most of the ownership expenses continue while the equipment is idle. These idle time costs must be precisely defined and accurately recorded to provide input to the burden cost computation and to supply management with product and capacity planning and scheduling decision criteria.

As an industry becomes more capital intensive, facility and equipment ownership costs and indirect operating expenses become a larger portion of the product cost as compared to the labor and material costs. As the high tech manufacturing system becomes more sophisticated and productive, the machining time cycle for each part and operation also shrinks. This makes each operation's cycle time, and the work station's "use rate" charge to the part, smaller. This tightening of the work elements makes the requirement for precise cost definition more critical. All of the costs of manufacturing must be included in the inventory valuation of the parts and the product.

The idle time costs must not be charged to the product! They are a cost of doing business and must be treated as a part of the general overhead of the enterprise.

D. HIGH TECH PRODUCTS REQUIRE PRECISION COSTS

The complexity of most consumer hard goods, military equipment, and industrial products is increasing very rapidly. Materials technology, in combination with the agendas for recycling, conservation, quality function deployment (QFD), and environmental protection, have raised the design standards for almost all products. The emphasis on total quality management (TQM) and the trend toward miniaturization have further complicated product design and manufacturing process requirements.

The mass-marketing capabilities of television advertising and chain stores, the market demand for quick response, and the increasing diversity of product models, require high-volume, and very flexible, manufacturing support. At the same time, the economic pressure to reduce inventory investment, and the application of group technology in manufacturing, have emphasized the need for standardization of parts and components and the coordination of product designs.

The very competitive cost/price trends and performance developments in high tech products are demonstrated by the concurrent technical development, declining cost, and diversified marketing of personal computers. Both their product designs and their high tech manufacturing processes have advanced at a rapidly accelerating rate. Their production volume has also increased greatly, and since the prices have declined sharply, one must assume that the manufacturing costs have been reduced.

The shortening idea/design to production time and reduced market lead time in the whole consumer electronics industry has forced all of industry to take a new look at their product development and manufacturing techniques.

A Look into the Future

It is obvious that rapid product development and short market introduction lead time requires accurate cost estimates to support the manufacturing and marketing decisions. It is also known that market forecasting is not a precise procedure, and that manufacturing volume projections based on these forecasts can be erroneous.

These are some reasons for using the volume independent time use of facility method of precision manufacturing costing instead of the traditional and conventional volume based distribution type costing techniques.

In a rapidly developing high tech environment, each design change can alter a product's manufacturing cost in several ways. A change in the materials used may modify the process and will usually affect material cost. A change in the design will usually require a revised process and result in a different operation cost. Different purchased components often have a price change. These cost changes can be cumulative, compensating, or a combination of these. As a result, the total product cost may or may not change.

It is essential to have detailed part and operations' cost data in order to estimate and predict the effect of proposed design changes or process modifications before a decision is made to implement them. The detailed element costs of the precision costing system will allow the planner and engineer to synthesize the product cost which will result from the proposed design or process changes. This will allow management to make an informed decision on a course of action.

In the rapidly changing high tech industries, whether in high volume production or in one of a kind operations, this precision costing capability is essential. These cost estimates must also be independent of anticipated sales volumes. The effect of sales volume predictions can be dealt with by using the precise costs as the basis for constructing a pro forma profit and loss statement, break even charts, marginal profit analysis, and enterprise performance projections.

Precise cost estimates must be developed before changes are made in order to avoid design and process actions and management decisions which could have disastrous cost and profit effects.

Management requires precise cost data as the basis for sound decisions in dealing with these business and engineering issues.

E. GLOBAL OPERATIONS REQUIRE PRECISE COSTING

The measurement and computation of manufacturing costs in an enterprise which operates only within a single national economy should be stable and precise, except for the effects of inflation or deflation on materials, overhead, and labor costs. If correctly defined and measured, the standard cycle times for the production operations will remain constant. The labor and work station time standards, and the materials' quantity standards should all be constant. In a stable economy, the standard unit cost of parts and products will also be stable.

Any cost variances from these defined standards should be attributable to methods changes or deviations in the cycle times and/or modifications in the cost of labor and materials. Labor and material costs will vary with inflation and wage rates. Changes in the work station "use rate" charges can also result from inflation or deflation of overhead expenses and taxes.

When the manufacturing economy becomes global, a new element enters the cost computation. The comparative cost calculation between countries is more complicated. In addition to differences in labor rates, overhead, and taxes, the rates of exchange between the countries' currencies can be a critical cost factor. The relative and fluctuating values of currencies are a key element in the computation of cost/price relationships and affect the product and component costs in international trade.

For example, let us assume that Japanese workers are paid at the stable rate of Y 1300 per hour. When the value of the Yen dropped from Y 130 to the dollar down to Y 103 to the dollar, the workers continued to make one hour of production in each Y 1300 hour. However, the dollar value of that production increased from $10.00 per hour to $12.62 on the world trade market. This is an increase of 26.2% with no change in the work cycle time, work method, or basic costs within Japan. Conversely, when the rate of exchange rises from Y 103 back up to Y 140 to the dollar, the international trading cost of the standard hour of work will go down to $9.28 for a 26.4% drop in dollar value, also with no change in the time or method of manufacture.

The fluctuation in rates of exchange is constant and global. It is a significant consideration in the prediction, definition, and control of the cost of manufacturing in a global market. It is a major cost management problem in an international manufacturing enterprise.

This rate of exchange problem is another valid reason for the detailed capture of all cost elements at every stage of the manufacturing process and in WIP inventory valuation. It is particularly important to compute precise standard costs in situations where the components and parts are moved across borders between factories of an international manufacturing company. If all of the costs are not precisely defined, there will be significant, and often cumulative errors in the international costing of the product.

In a multinational operation, the raw materials and WIP inventory values should be defined and standardized globally on the basis of a common corporate currency (usually the U.S. dollar) during the manufacturing process. This will allow a realistic comparison of cost performance among the various locations. The rate of exchange adjustments should be made in the financial analysis and not in the manufacturing cost determination.

A Look into the Future

Production cost variances resulting from deviations from predicted operating standards, such as changes in cycle time or labor cost should be treated as burden in the same manner as discussed above. Changes in the rate of exchange between international segments of the enterprise are also variances. Their impact should be treated as corporate overhead and be fed back into the product cost through the space "use rate" of the using or buying operation. This will allow a more realistic basis for comparing source pricing and manufacturing costs between locations and countries.

The basic precision of the cost definition procedure in each segment of a global manufacturing system is an essential element of the international cost management capability of the enterprise. By establishing a basic engineered cost definition for each operation and part, and by using a corporate-wide base currency to define the cost of each part and process, management can see and measure the performance variations. The financial impact of currency fluctuations should be applied to the total cost of the parts or products as a part of the inventory valuation or pricing procedure and as a profit and loss computation.

Global operations with a variety of labor wage rates, materials costs, taxes, and facility overhead expenses require precise base cost data in order to provide valid and comparable standards for comparison and decision making. Inadequate detail, when coupled with variations in the rates of exchange, can lead to significant cost computation errors. This can deceive management into assuming cost advantages when none exist, or ignoring cost savings because of the failure of the cost system to identify and highlight them. Precision costing is an essential international management tool.

F. THE OLD WAYS ARE INADEQUATE—NEW PRECISION COSTING IS REQUIRED

Many people believe that the only sure thing in life is continual change. As in the Industrial Revolution of the past, continual change is a certainty in today's high technology revolution. Change will continue in the global market economy of the world's industrial future.

Every day, and often every hour of every day, a new technique, product, or process is born. New markets are continually developing throughout the globe. The world's political and industrial economic systems are in a continuous state of development. These changes are usually gradual, subtle, and often hard to anticipate. They are occasionally rapid and revolutionary.

Even the most stable of industries must cope with these continuing changes. Failure to change will cause loss of market share or displacement by new technologies and new or replacement products. It has been said that the design of an airplane is not completed until the weight of the change order documents equals the weight of the airplane. This could be true in the manufacture of most high tech products.

The measurement of change, and its impact on product and performance, is a primary requirement of management and design and process engineering. The measurement of change has two primary objectives.

1. Management must define and measure variances from business expectations and manufacturing plans in order to appropriately and competitively respond by implementing corrective control actions or by producing innovative and competitive counter measures.
2. Engineering must define, and precisely estimate the cost effect of each product and process design and design change before its implementation and production. Engineering must identify and measure the performance and cost effect of any variances from the predicted product or process performance. Engineering must define the cost effect of each new design feature and process procedure during production and, after the product is produced in the factory, make the appropriate, responsive, and competitive corrections in the product design and manufacturing process.

Before the availability of computers, achievement of precise measurements of performance was inhibited by the cost of management and engineering staffing, the limited ability to collect precise data, and the time-consuming need for manual computations. These obstacles made the evaluation process and responsive correction slow and often inaccurate. The analysis and evaluation time factor often caused the manufacture of obsolete and unsalable inventory. In order to cope with these limitations, analysts often resorted to sampling techniques and used percentages to interpret the data.

The basic cost data was usually produced by using the conventional distribution type accounting procedures. In those procedures, the application of overhead and burden costs to the product was tied to the variable production volume. The cost of using the facilities and machinery was not focused on the specific part or product's operation. It was often assigned to the product cost as a percentage of labor or material cost. The general and administrative cost (G & A) was usually omitted from the product cost computation.

These generally accepted accounting practices do not result in a correct and true definition of the cost of the product or the operations used in its manufacture. The apparent cost of the product varies with production volume. If the overhead distribution is tied to a percentage of material or labor cost, the apparent overhead also varies with wage fluctuations and material prices.

These old ways produce varying and erroneous product cost estimates. A *new approach* to manufacturing costing is required! It is essential to separate the cost of the product and process from the volume of production. The

A Look into the Future

"Precision Manufacturing Costing" approach is designed to overcome these approximations, errors, and inadequacies by focusing the costing procedure on the individual operations and parts.

By using a total absorption approach to the costing of operations, it is possible to cumulatively assign the correct share of indirect or burden, overhead, and G & A manufacturing costs to each part or component on a level-by-level work-in-process, inventory valuation basis. By distributing these costs to the product on the basis of a time use of facility cost or "use rate" for each specific work station and its equipment, and for each operation in the process, it is possible to cost the product at any point in the process without reference to the volume of production in the plant. This results in a "pure" assignment of the proper share of indirect expenses to the product or part for each operation.

By adding the time based total cost of direct labor (including fringes) for the operation, and the adjusted standard material cost for the part, to the work station's operation "use rate" for the part, one can develop a "true cost" of each operation, part, and product without consideration of the volume produced. This cost is particularly important in the comparative evaluation of performance at different plants and at different times in a seasonal annual production cycle. In this procedure, the operation costs will not vary with production volume.

As the manufacturing environment becomes less labor intensive and more high tech and capital intensive, the error factors inherent to the "generally accepted accounting practice" method of distribution cost accounting become more onerous. The future CIM factory with few people and little direct labor needs to focus the cost analysis on the use of specific production machinery and systems.

The time use of facility approach accomplishes this objective in a manner which allows detailed analysis of both the product and process design and manufacturing performance. At the same time it isolates the manufacturing costs from non-manufacturing business expenses.

With the availability of the computer and the use of data base software, all of the accounting details are available. The tax accountant, financial manager, banker, and general manager can each have his own presentation format from the same data base without depending upon the factory cost information which may not balance with the financial report.

At the same time, the design engineer, process engineer, and factory manager will have the precise and focused data needed for analysis of design alternatives, different processes, and daily operations.

The procedures presented in this book are capable of dealing with the needs of the capital intensive CIM factory of the future, the multiplant and multinational corporation, the location based variations in labor costs and rates of exchange, and the requirement for precision definition of the cost of manufacturing.

These are the proper cost accounting tools for the future world of capital intensive manufacturing. They can be applied to:

- Computer integrated manufacturing (CIM)
- Flexible manufacturing systems (FMS)
- High-volume, continuous hard automation
- Job shop manufacturing operations
- High-volume, serialized piece part production
- Continuous process systems (chemical, paper, food)
- Warehousing and physical distribution operations

This book offers a "Precision Manufacturing Costing" system for manufacturing cost management, product design, cost estimating, inventory valuation, and distribution costing, for use in the capital-intensive and multinational high technology industrial future!

The old ways are obsolete and inadequate for the twenty-first century.

Discussion Questions

The following conditions apply to all problems unless otherwise stated:

- The depreciation rates are straight line 20 years for buildings and 10 years for machinery.
- The ownership cost for inventory is 20% per year.
- The insurance rate is 5% of building and machinery investment.
- The income tax rate is 38% of a taxable profit of 12% based on annual sales of 2.5 times the invested capital.
- The plant building costs $50 per sq ft and operates on a 40 hour week, 52 weeks per year (2080 hours) with individual two week vacations and six paid shut down holidays, but no plant shut down.

(Correct answers are indicated in **bold**.)

1. The indent code in the SIMSCODER matrix part numbering system shows **1**. Level of completion of the part, 2. Functional classification of the part, 3. Product shape configuration, 4. All of these, 5. None of these.
2. General and administrative expense (G & A) is computed on the basis of the value of the machinery and space in a work station. **1**. True, 2. False.
3. The 50% premium paid the labor force for working more than 40 hours in a week is 1. Required by law, 2. The cost of scheduling, 3. "For nothing" cost, 4. None of these, **5**. All of these.
4. The only fixed bases for cost allocation in manufacturing and warehousing are 1. Labor and materials, **2**. Time and space, 3. Dollars, 4. None of these, 5. All of these.

5. The invested capital in a work station is represented by the original purchase value of the machinery and the building space occupied. **1.** True, 2. False.
6. Manufacturing burden expense includes such items as **1.** Toilet paper and floor mops, 2. The salary of the President's secretary, 3. Advertisements in magazines, 4. None of these, 5. All of these.
7. The President and his staff earn salaries and have expense accounts which total $2,000,000 per year. The company has invested $4,000,000 in machinery, $3,000,000 in buildings, and has $1,000,000 in WIP inventory. The company operates on a 40 hour per week basis with six holidays and a two week vacation shutdown. The buildings cost $50/sq ft to build. The capital invested in a particular work station and its space is $300,000. The G & A element of the work station use rate is 1. $41.19/$/Hr, 2. $0.0001463/Hr, 3. $42.84/Hr, **4.** $43.89/Hr, 5. None of these.
8. A plant has an area of 100,000 sq ft and costs $50/sq ft to build. The manufacturing management staff earns a total of $1,500,000 in salaries. Miscellaneous factory supplies and overhead expenses including insurance, interest, taxes, and utilities are $500,000 per year. The burden element of the hourly space "use rate" of a 400 square foot work station which operates with two week vacations and 6 holidays is: 1. $3.36, 2. $4.32, 3. $4.50, **4.** $4.61, 5. None of these.
9. A part is produced on a mill/turn lathe which costs $250,000 and operates for 8 hours per day in a plant which has six paid idle holidays and no annual shutdown. Production is at the rate of 2 minutes per piece. The work station occupies 100 square feet of $50/sq ft space. The burden labor and burden material rate is $0.95/hour for the work station. The G & A charge rate is $0.000025/$/hr. The use rate charge for the work station is 1. $12.019/hr, 2. $18.51/hr, **3.** $19.75/hr, 4. $18.678/hr, 5. $6.375, 6. None of these
10. Using the data from Problem 9., and adding $1.05 per piece for material, the cost of one piece without labor is 1. $7.425, 2. $20.80, 3. $13.353, **4.** $1.708, 5. None of these.
11. Using distribution type accounting principles, what is the cost of a product (not including direct labor) when the total annual burden expense is $6,000,000, the machinery depreciation is $200,000 per year, the material cost is $2.00, and the constant production rate is 50 pieces per hour? **1.** $63.02, 2. $61.02, 3. $59.61, 4. $61.61, 5. None of these.

Discussion Questions

12. The burden costs in a manufacturing plant do not include 1. The cost of inspectors, 2. The wages of the forklift operators, 3. The salary of the Treasurer, 4. All of these, 5. None of these.
13. G & A expenses are 1. General and administrative expense, 2. The cost of the President and his staff, 3. The cost of the executive offices, **4**. All of these, 5. None of these.
14. Standard material cost is defined as 1. The net shape of the finished part, **2**. The original piece of material needed to start making the part, 3. The purchased material used in the product, 4. None of these, 5. All of these.
15. The only material of production which earns money when it is idle is 1. Steel, 2. Purchased parts, **3**. Money, 4. Brass, 5. None of these.
16. The manufacturing bill of materials is designed to show **1**. The assembly structure of the product, 2. The technical structure of the product, 3. The manufactured parts and materials in the product, 4. None of these, 5. All of these.
17. A matrix part numbering system can help engineering by providing access through 1. Part function, 2. Bill of materials explosions, 3. Where used analysis, 3. None of these, **5**. All of these.
18. Parts standardization requires 1. The same part number on a part regardless of its location in the bill of materials, 2. Common definition or description of the part in all of its applications, 3. Identical part design in every application, 4. None of these, **5**. All of these.
19. Group technology is important to cost system design because it 1. Helps to minimize the number and variety of setups and processes, 2. Helps to minimize the complexity of tooling, 3. Reduces the cost of setups, 4. None of these, **5**. All of these.
20. Value engineering is 1. A pricing procedure, **2**. The analysis of products to reduce cost without reducing function or quality, 3. An accounting procedure which uses engineering, 4. None of these, 5. All of these.
21. Total absorption standard costing differs from conventional standard costing because it 1. Fully accounts for all indirect cost elements at every step in the process, 2. Fully defines the cost of work-in-process inventories, 3. Does not charge unused facilities into the cost of the product, **4**. All of these, 5. None of these.
22. In the application of "use rate" and total absorption cost systems, the cost of unused facilities is charged to 1. The product, **2**. The profit and loss statement, 3. The factory, 4. None of these, 5. All of these.

Discussion Questions

23. Fringe benefits do not include 1. Health insurance premiums, 2. Vacation and holiday pay, 3. Overtime bonuses, 4. None of these, 5. All of these.
24. The use of percentages in computing expected scrap rates is realistic at all levels of production. 1. True, 2. False.
25. The learning curve applies only when 1. A process is highly mechanized, 2. A process is labor intensive, 3. In very automated plants, 4. None of these, 5. All of these.
26. The "nest and gain" factor is the material usage effect of 1. Labor productivity, 2. Scheduling of products in the system, 3. Purchasing negotiation, 4. None of these, 5. All of these.
27. Generally accepted accounting practice is normally 1. Tax law oriented, 2. Based on distribution type accounting, 3. Based on percentage application of indirect expenses, 4. None of these, 5. All of these.
28. The application of total absorption standard costing often requires the use of two sets of accounting records and is difficult to apply without a computer. 1. True, 2. False.
29. The process of estimating product cost is easiest with total absorption standard costing and matrix coded numerical bills of materials because 1. Information retrieval is easier, 2. Each part is fully costed in the inventory data base, 3. The bills of materials can be fully costed from prior data, 4. All of these, 5. None of these.
30. In the application of group technology codes, the code should address 1. Part geometry, 2. Part function, 3. Process and material, 4. All of these, 5. None of these
31. The work station "use rate" for a CNC milling center is $60/hour. The production rate for Part A is 3 pieces per hour with a material cost of $2.50 and a labor rate of $15.00 per hour. The cost per part is 1. $25.00, 2. $20.83, 3. $25.83, 4. $27.50, 5. None of these.
32. An emergency part order is made on the milling center used in problem No. 31. The job is started at 2:00 PM on Friday and completed at 9:00 PM in the evening. The machinist has worked eight hours every day and normally quits at 4:30 PM. The "for nothing" cost (without fringes) of scheduling this job for completion after a 30 minute unpaid supper break is 1. $90.00, 2. $30.00, 3. $60.00, 4. $45.00, 5. None of these.
33. The identity of the process should be included in the code structure for group technology. 1. True, 2. False.
34. Configuration control is the term used to describe the process of dealing with 1. The shape of a part, 2. The organization of the

Discussion Questions

engineering department, 3. Engineering changes in a part or product, 4. None of these, 5. All of these.

35. The difficulty factor is a term used in the calculation of 1. Process time, **2**. Scrap allowances, 3. Tooling needs, 4. None of these, 5. All of these.
36. In the absence of statistical data, it is best to anticipate scrap losses by adding 1. A fixed percentage of the scheduled batch or lot, 2. A fixed quantity to the batch or lot, **3**. A quantity equal to the square root of the batch or lot, 4. None of these, 5. All of these.
37. The financial staffs of many firms have two key executives who are 1. President and Treasurer, **2**. Treasurer and Controller, 3. President and Secretary, 4. President and Controller, 5. None of these.
38. Work measurement is a basic element of precision cost accounting **1**. True, 2. False.
39. MOST is a term which describes 1. Modern operational standards techniques, **2**. Maynard operations sequence technique, 3. Management organizational system technique, 4. None of these, 5. All of these.
40. MTM is a stop watch time study technique 1. True, **2**. False.
41. Work sampling is a predetermined time study technique 1. True, **2**. False.
42. If a worker is assigned to operate a machine which is more complex and requires a higher skill rating than his/her normal job, the effect on the cost system will be **1**. A plus variance in the cost of the part being made, 2. A plus variance in the payroll account and no change in the part cost, 3. No change in the worker's pay and a negative variance in the part cost, 4. None of these, 5. All of these.
43. When a part is made on a higher cost machine than normally scheduled, the cost effect will be 1. A plus variance in the machine use rate only, **2**. A plus variance in both the machine use rate and the labor rate, 3. No change in the elements of the calculated cost, 4. None of these, 5. All of these.
44. In assembly operations, the work station "use rate" is based on ____ cost elements as compared to machine work stations. 1. Different, **2**. The same, 3. Labor only, 4. None of these, 5. All of these.
45. In designing a computer program for a precision cost system it is best to 1. Start from scratch and design a new data base program, **2**. Build around an existing data base program such as dBaseIII+ and its competitors, 3. Design a for-purpose mathematical algorithm and system, 4. None of these, 5. All of these.

324   Discussion Questions

46. The precision cost computer system must interface with other computer programs for applying 1. MRP, 2. Sales, 3. CAD/CAM, 4. None of these, **5**. All of these.
47. In states which have an inventory ownership tax, as in Ohio, the use of total absorption standard costing could _____ the annual inventory tax. 1. Decrease, **2**. Increase, 3. Not effect, 4. None of these, 5. All of these.
48. The application of use rate concepts and the time use of facility cost accounting method will 1. Allow responsibility accounting to be more realistic, 2. Permit allocation of "blame" for unused facility and profit loss to the sales function, 3. Provide a basis for more accurate and volume independent product cost estimating, 4. None of these, **5**. All of these.
49. Application of total absorption standard costing with its straight line depreciation base will usually require the company to 1. Use a second set of accounts for financial and tax management, 2. Use a precise data collection system and a data base type computer program, 3. Prepare both an operational and a financial type set of performance reports, 4. None of these, **5**. All of these.
50. The way that an executive or engineer dresses and presents himself in business and the appearance of the documents presented will have a major impact on employer management's perception of, and confidence in the professionalism and validity of the individual's work. **1**. True, 2. False.

References

Ashtan, James E. and Neal Holmlun. 1988. Relevant Managerial Accounting in the Job Shop Environment. *Manufacturing Review*, Vol. 1, No. 4, pp. 230–235.
Baker, Eugene F. 1985. *Industry Shows Its Stripe*. New York: American Management Association.
Blake, Michael O. 1989. Capacity and Cost Analysis for Cell Manufacturing System. *Manufacturing Review*, Vol. 2, No. 3, pp. 214–221.
Blank, Leland T., and Anthony J. Tarquin, 1989. *Engineering Economy*, Third Edition. New York: McGraw-Hill.
Boer, Germain. 1987. *Classifying and Coding for Accounting Operations*. Montvale: National Association of Accountants.
Brathwaite, Kenneth S. 1989. *Systems Design in Database Environment*. New York: Multiscience Press.
Brimson, James A. 1988. Bringing Cost Management Up to Date. *Manufacturing Engineering*, June, pp. 49–51.
Calvasino, Richard V. and Eugene J. Calvasino. 1984. Standard Costing Games that Managers Play. *Management Accounting*, March, pp. 49–50.
Cheatham, Carole. 1990. Updating Standard Costing Systems. *Journal of Accountancy*, Dec., pp. 57–60.
Collins, Frank and Michael L. Werrner. 1990. Focus on Industry. *Journal of Accounting*, June, pp. 131–132.
Cooper, Robin and Robert, and S. Kaplan. 1988. Measure Costs Right: Make the Right Decisions. *Harvard Business Review*, Sept./Oct., pp. 96–103.
Dhavale, Dileep G. 1988. Indirect Costs Take on Greater Importance, Require New Accounting Methods with CIM. *Industrial Engineering*, July, pp. 41–43.
Drop, Rob. 1990. *Working with dBASE Languages*. Chichester, West Sussex: Ellis Horwood.

Elier, Riler G., Walter K. Goletz, and Daniel P. Keegan. 1982. Is Your Cost Accounting Up to Date? *Harvard Business Review*, July/Aug., pp. 133-139.

Emig, James M. and Matthew Mazeffa. 1990. Victims of Cost Accounting. *The National Public Accountant*, April, pp. 30-34.

Fabrycky, Wolter J. and Blanchard, Benjamin S. 1991. *Life-Cycle Cost and Economic Analysis*, Englewood Cliffs, N.J.: Prentice-Hall.

Ferrara, William L. 1990. More Questions Than Answers. *Management Accounting*, Oct., pp. 48-52.

Fremgen, James M. 1972. *Accounting for Managerial Analysis*, Homewood. Ill. Richard D. Irwin.

Gilligan, Brian P. 1990. Traditional Cost Accounting Needs Some Adjustments... as Easy as ABC. *Industrial Engineering*, April, pp. 34-38.

Herbert, Popper. 1970. *Modern Cost Engineering Techniques*. New York: McGraw-Hill.

Hodson, William K. 1992. *Maynard's Industrial Engineering Handbook*, Fourth Edition. New York: McGraw-Hill. Howell, Robert A. and Stephen R. Soucy, 1987. Cost Accounting in the New Manufacturing Environment. *Management Accounting*, Aug., pp. 42-48.

Humphreys, Kenneth K. and Paul Wellman. 1987. *Basic Cost Engineering*, 2nd ed. New York: Marcel Dekker.

Johnson, H. Thomas. 1989. Managing Costs: An Outmoded Philosophy. *Manufacturing Engineering*, May, pp.

Johnson, H. Thomas. 1991. Activity-Based Management: Past, Present, and Future. *The Engineering Economist*, Vol 36-, No. 3, pp.

Kaplan, Robert S. 1987. The Rise and Fall of Management Accounting. *Management Accounting*, January, pp.

Kaplan, Robert S. 1988. One Cost System Isn't Enough. *Harvard Business Review*, Jan./Feb., 61-66.

Keys, David E. 1986. Six Problems in Accounting for N/C Machines. *Management Accounting*, Nov., pp. 38-47.

Knecht, Ken. 1986. *Practical Paradox*. Blue Ridge Summit: Tab Books.

McFadden, Fred R. and Jeffery A. Hoffer. 1991. *Database Management*. 3rd ed. Redwood City: The Benjamin/Cummings Publishing Company.

Michaels, Jack V. and William P. Wood. 1989. *Design to Cost*. New York: John Wiley & Sons.

Mittar, Sitansu S. 1991. *Principles of Relational Database Systems*. Englewood Cliffs: Prentice-Hall.

Montevale, 1988. *Adapting Management Accounting Practice to an Advanced Manufacturing Environment*. National Association of Accountants.

Ostrenga, Michael R. 1990. Activities: the Focal Point of Total Cost Management. *Management Accounting*, Feb., pp. 42-49.

Ostwald, Philip F. and Patrick J. Toole. 1978. IE's and Cost Estimating. *Industrial Engineering*, Feb., pp. 40-42.

Peavey, Dennis E. 1990. It's Time for a Change. *Management Accounting*, Feb., pp. 53-56.

Petreley, Nicholas, Alex Khologhli, and David Chalmers. 1989. Gang of 4 Multiuser Databases. *InfoWorld*, April, pp. 66-67, 70-72, 74, 78, 80, 82-83, 86-87, 89-90.

References

Petreley, Nicholas, Zoreh Banapoor, and Linda Slovick. 1990. Analyzing Rational Databases. *Infoworld*, Jan., pp. 51-54, 56-61, 64, 66-68.
Poor, Alfred. 1989. The Art of Simultaneous Action. *PC Magazine*, Sept., pp. 105-112, 116, 118-121, 126, 128, 133-140, 143-146.
Rafish, Norm. 1991. How Much Does That Product Really Cost? *Management Accounting*, March, pp. 36-39.
Ricardo, Catherine. 1990. *Database Systems: Principles, Design, and Implementation.* New York: Macmillan Publishing Company.
Romano, Patrick L. 1990. Where is Cost Management Going? *Management Accounting*, Aug., pp. 53-56.
Roth, Harold P. and A. Faye Borthick. 1989. Getting Closer to Real Product Costs. *Management Accounting*, May, pp. 28-33.
Salvendy, Gavriel. 1992. *Handbook of Industrial Engineering,* Second Edition. New York: John Wiley & Sons.
Schwarsback, Henry R. and Richard G. Vangermeersch. 1983. Who Should Account for the 4th Cost of Manufacturing? *Management Accounting*, July, pp. 24-28.
Seed, Allen H. 1990. Improving Cost Management. *Management Accounting*, Feb., pp. 27-30.
Sharp, Douglas and Linda F. Christensen. 1991. A New View of Activity-Based Costing. *Management Accounting*, Sept., pp. 32-34.
Sims, E. Ralph, Jr. 1967. *Euphonious Coding.* Cambridge, Mass: Management Center of Cambridge.
Sims, E. Ralph, Jr. 1968. *Precision Costing of Manufacturing Operations.* Paper No. 680575, Ann Arbor, Mich: Society of Automotive Engineers.
Sims, E. Ralph, Jr. 1968. *Planning and Managing Materials Flow.* Boston: Industrial Education Institute & Farnsworth Publishing.
Sinclair, Neil D. 1989. *Estimating for Abnormal Conditions.* New York: Industrial Press.
Sourwine, Darrel A. 1989. Does Your System Need Repair? *Management Accounting,* Feb., pp. 32-36.
Stalk, George, Jr. 1988. Time—The Next Source of Competitive Advantage, *Harvard Business Review,* July/Aug. pp.
Steiner, Henry Malcolm. 1992. *Engineering Economic Principles* New York: McGraw-Hill.
Sutton, Sharon G. 1991. A New Age of Accounting. *Production and Inventory Management Journal*, 1st Qtr, pp. 72-75.
Tatikonda, Lakshmi U. and Rao J. Tatikonda. 1991. Overhead Cost Control—Through Allocation or Elimination. *Production and Inventory Management Journal*, 1st Qtr, pp. 37-41.
Thuesen, G.J. and W. J. Fabrycky. 1993. *Engineering Economy*, Eighth Edition. Englewood Cliffs, NJ: Prentice Hall.
Tompkins, James A. and Dale Harmelink. 1994. *The Distribution Management Handbook*, New York: McGraw-Hill.
Tompkins, James A. and Jerry D. Smith. 1988 *The Warehouse Management Handbook,* New York: McGraw-Hill.
Vasta, James. 1985. *Understanding Database Management Systems*, Belmont: Wadsworth.

Vernon, Ivan R. 1968. *Realistic Cost Estimating for Manufacturing*. Dearborn, Mich: American Society of Tool and Manufacturing Engineers.

Wolf, Warren G. 1982. Developing a Cost System for Today's Decision Making. *Management Accounting*, Dec., pp. 19-23.

Woods, Michael D. 1989. New Manufacturing Practices-New Accounting Practices. *Production and Inventory Management Journal*, 4th Qtr, pp. 7-11.

Zandin, Kjell B. 1980. *MOST Work Measurement Systems*. New York: Marcel Dekker.

Index

Accounting
  accrual method 15, 17
  cash method 15
  distribution 21
  life cycle 10
  percentage cost 21
  responsibility 13
  technology 10
Activity Based Costing 9, 22, 53, 204, 220, 285, 574
American National Standards Institute 129
ANSI/ASQC Q90 129
Assembly operations 276
Automation
  hard automation 5
Average Outgoing Quality Level (AOQL) 133
Average time standards 39

Bill of materials
  engineering (EBoM) 33, 95
  manufacturing (MBoM) 33
  parts list 33

Bonuses
  fringe benefits 73, 158, 260
  group bonus plans 157
  holiday work 260
  management 74
  night shift 260
  profit sharing 157
  shared gain plans 157
Book value 191
Buildings 85, 193
Burden 168
  buildings 191
  labor 76
  materials 40, 42, 77, 140

Call in pay 158
Capital account number 171
Capital investment 198
  facility investment 191
Common denominator 69, 224, 281, 300
Computer 6
  Computer Aided Manufacturing International (CAM-I) 8

[Computer]
  Computer Integrated Manufacturing (CIM) 1
  Computer Numerical Control (CNC) 296
Contingent materials 40, 42, 140, 173, 186
Cost Accounting Standards Board 8
Cost variances 289
  equipment variances 295
  labor variances 293
  material cost variances 299
  process variances 297
  product cost variances 299
  warehouse cost variances 300
Costing
  Activity Based Costing (ABC) 9
  direct costing 21
  equipment costing 42, 84
  facilities 42
  machine labor costing 9
  precision costing 47
  precision engineered costing 12
  process costing 9
  productive hour rate costing 10
  scrap costing 80, 134
  standard costing 11, 23, 41, 100
Critical Path Method (CPM) 28
Cube 224, 226

Depreciation 25, 43, 67, 168, 192
  MACRS depreciation 24, 25, 135
  straight line depreciation 136
  tax life depreciation 68, 85
Design 56, 156
  design to cost 217
  design standard material 138, 158, 290
Direct labor 76, 235
Direct materials 39, 138
Distributive cost accounting 33

Dollar density 224
Drop off 264

Economic Order Quantity (EOQ) 131
Eli Whitney 35
Engineering
  concurrent engineering 43
  design engineering 56, 95, 217
  engineering changes 95, 116
  manufacturing engineering 95
Engineering bill of materials (EBoM) 95
Estimating costs 42, 84, 167
Estimating 56
European Economic Community (EEC) 129

Facility capital investment 191
Factory overhead 184
Financial management 128
Flexible Manufacturing Systems (FMS) 1, 4, 11, 296
For nothing cost 74, 156, 260
Fringe benefits 73, 155, 158, 165, 260
  health and hospitalization insurance 260
  holiday work bonus 260
  military leave 260
  social security taxes 259
  unemployment compensation 260

General and administrative expense (G & A) 59, 82, 86, 197, 200, 202, 263
Generally accepted accounting practices 2
Gilbreth, Frank and Lillian 35
Group bonus plans 157
Group technology 214

Handling class 238

Index

Health and hospitalization insurance 260
Holiday work bonus 260

Idle capacity 174
Incentive compensation 75, 156
  group bonus plans 157
  management bonus 74
  measured day work 157
  negative incentives 157
  piece work 156
  profit sharing 157
  shared gain 157
  standard hour incentives 156
Indented manufacturing bill of materials 276
Indirect labor 76, 183
Indirect materials 40, 42, 140
Industrial engineering 31, 40
Industrial revolution 1, 5, 305
Information system 93, 96
Insurance 187
Internal Revenue Service (IRS) 2, 84
International Standards Organization (ISO) 129
ISO 9000 129
Inventory 2, 27, 71, 78, 140, 143, 200, 208, 233
  inventory costing 27, 140, 151
  work-in-process 5, 148
Invested capital 87, 201

KanBan 8, 102

Labor 83
  burden labor 76
  direct labor 76, 235
  indirect labor 76, 183
  labor standards 159, 160
Land 192
Lean production 2

Level-by-level 12, 31, 53, 57, 59, 70, 105, 117, 119, 142, 290

Machinery 193
MACRS 24, 25, 135
Maintenance 169, 172, 193
Management Information System (MIS) 94
Manufacturing Bill of Materials (MBoM) 15, 33, 53, 57, 104, 146, 290
Manufacturing operation costs 272
Manufacturing operations 182
Materials cost 186, 250
Materials Requirements Planning (MRP) 78, 145
Materials handling 223, 229
Materials management 78, 145
Materials standards 39
Maynard, Stegmerton, and Schwab 35
Measured day work 157
Military leave 260
Modified Accelerated Cost Recovery System (MACRS) 24, 25, 68, 135

Negative incentives 157
Nest and gain 41, 50, 79, 136
Net operating space 52, 67
Night shift bonus 260

Part number coding 290
Payroll 155
PERT 28
Piece work incentives 156
Precision costing 4, 12, 24, 85, 119, 209, 286, 312
Price 13, 219
Procurement 129
Product cost budget 28, 46, 89, 152, 178

Production volume 49, 208
Profit sharing 157

Quality 133
   Total Quality Management (TQM) 133
   Quality Function Deployment (QFD) 93

Rating 37
Rent 192
Responsibility accounting 219

Salary 155
Scientific management 34
Scrap cost 80, 134
Securities and Exchange Commission (SEC) 2
Shared gain plans 157
SIMSCODER 101, 108, 210
Social Security taxes 259
Space "use rate," 52, 67, 85
Standard time 37, 161
Storage 223

Tax life 68, 168
Taxes 25, 84, 187
Taylor, Fredrick W. 35
Therblig 35
Time standards 35, 160
   MOST = Maynard Operations Sequence Technique 36

[Time standards]
   MTM = Methods Time Measurement 24, 35
   Pre-determined Work Time Standards 35
Time use of facilities 35, 41
Tooling 173
Total absorption standard costing 119, 286

Unemployment compensation 260
Union contracts 73, 158
Use rate 12, 50, 59, 60, 174, 177, 195, 202, 205, 223, 226, 261, 290
   equipment use rates 44, 53, 86
   general and administrative expense (G & A) use rate 88
   space use rates 52, 67, 85
   work station use rates 52, 62, 261
Utilities 173, 194

Warehouse operations 181, 223
Wilson formula 131
Work-in-process 5, 148
Work measurement 32, 35, 62
   work sampling 37
Work station use rates 52, 62, 261
Worker's compensation insurance 259

Zandin, Kjell B. 36